I0479621

Inverse Gravity Cosmology

The mathematics, success, and failure of a journey
in theoretical physics

Edward Alexander Walker

Illustrators:
Edward Alexander Walker
Angy Angrand

INVERSE GRAVITY COSMOLOGY
THE MATHEMATICS, SUCCESS, AND FAILURE
OF A JOURNEY IN THEORETICAL PHYSICS

iUniverse books may be ordered through booksellers or by contacting:

iUniverse
1663 Liberty Drive
Bloomington, IN 47403
www.iuniverse.com
844-349-9409

Because of the dynamic nature of the Internet, any web addresses or links contained in this book may have changed since publication and may no longer be valid. The views expressed in this work are solely those of the author and do not necessarily reflect the views of the publisher, and the publisher hereby disclaims any responsibility for them.

Any people depicted in stock imagery provided by Getty Images are models, and such images are being used for illustrative purposes only. Certain stock imagery © Getty Images.

ISBN: 978-1-6632-5434-4 (sc)
ISBN: 978-1-6632-5435-1 (e)

Library of Congress Control Number: 2024900076

Print information available on the last page.

iUniverse rev. date: 05/01/2024

For Lola

(Lola Mae Roberts — November 6, 1932–March 8, 2023)

Table of Contents

Preface

The inverse gravity cosmological concept is a hypothesized explanation of cosmological expansion. This book presents the integration of the theoretical concept to established cosmological notions and mathematics accepted by the physics community and based on observations. Furthermore, the inverse gravity concept is applied to the quantum realm, describing ideas such as linearized gravity and forms of dark matter such as weakly interacting massive particles which also incorporate quantum field theory. Lastly, mistakes in previous publications as well as new assertions in the theoretical concept are discussed.

Chapter 1

The Journey and Development of the Inverse Gravity Hypothesis and Concept

"My thoughts are not your thoughts, neither are your ways my ways, saith the LORD. For as the heavens are higher than the earth, so are my ways higher than your ways, and my thoughts than your thoughts" (Isaiah 55:8–9). The unfathomable mind of God as shown by his creation is an inconceivable fraction of why he deserves our glory and adoration.

The splendor and majesty of God's creation through Christ Jesus is amazingly complex and shows his invisible qualities, *"For the invisible things of him from the creation of the world are clearly seen, being understood by the things that are made, even his eternal power and Godhead"* (Romans 1:20). Thus, my inspiration and intrigue with Father Jehovah's universe is my motivation to exploring new directions in physics, thereby exploring and uncovering an infinitesimally small fraction of the mind of God who is our Father Christ Jesus.

We are dead in our sins and separated from God eternally; however, God loves us, God loves you! He sent his son to die for you! He made him who had no sin to become sin for us! (2 Corinthians 5:21). Jesus is the only way (John 14:6). So if you have not accepted Christ as your lord and savior, then today repent of your sins, confess with your mouth, and believe in your heart that Jesus is lord, and you will be saved, in the family of God, and connected to him eternally (Romans 10:9).

+

The development of the inverse gravity cosmological concept has been a journey of ascension, discovery, excitement, failure, and disappointment for me as a physics researcher. Despite being a culmination of all these things, the journey has been both enlightening and enriching to my growth as a physicist. Hence, the purpose of this book is to express my folly in this endeavor to gain an understanding of the universe and to thoroughly elucidate the progress of the theoretical inverse gravity cosmological concept and the integration of its aspects and applications to cosmology and beyond. Therefore, this endeavor has a purpose motivated by discovery and intrigue. I must include a caveat that the expressions, equations, formulations, derivations, and explanations are mathematically devout. This is not a book for the lay person, it is quite arcane and esoteric to those in the field.

My Journey

As a child, I wanted to be an astronaut. On the clear nights of south Florida in the late 1980s, I would look up at the stars and dream of traveling to them like all dorks do. And yes, I am a Trekkie! (a *Star Trek* fan). I was always in awe at the sight of God's universe. Images of stars and galaxies always amazed me. I loved everything from space travel, aerospace, and astronomy to cosmology, science fiction, marine biology, and more. I had an overall affinity for science! However, I did have my issues with learning as a child which I will revisit shortly.

I used to build model rockets and airplanes with my father. I began flight lessons at the age of twelve although I've never completed my private pilot's license (at least at the time of writing this book). With my love of aviation, I would try to find unique ways to build airplanes; aerospace engineer Burt Rutan was an inspiration to me as a child. At Sea Castle Elementary School, I would see a Rutan VariEz (an aircraft with a canard configuration designed by the great Burt Rutan) landing and taking off at North Perry airport in Pembroke Pines, Florida. Furthermore, I would read *Flying* and *Kit Plane* magazines, which further cultivated my desire to use my own creativity to construct and design especially airplanes.

I would occasionally apply my creativity in other areas of my life; I sketched and did other forms of artistic expression. As a child, I became absolutely fascinated with making my designs fly; our neighbors enjoyed watching both my successes and failures, almost betting on which one of my designs was airworthy. My learning process became one of creativity and trial and error.

As I mentioned, I had issues with learning in formal settings from a very young age; comprehension and understanding of any topic or subject was met with barriers. In my late twenties, my mother revealed that I was diagnosed with an attention deficit disorder. Mastery of anything took twice as long and double the effort of my peers. I had to truly learn the value of work and discipline to succeed in academics (which is the case even if you are a precocious individual). I later learned that I am autodidactic.

Moreover, for a kid with an affinity for the sciences, it was compulsory that I had a propensity for mathematics. Needless to say, from childhood, my love for and relationship with mathematics were tenuous. My grandfather, Edward Walker II, was a professor of mechanical engineering, and he would visit us once or twice a year from Detroit, Michigan. My grandfather would always put an emphasis on the importance of mathematics and how it applied to my interest, although I could not see the correlation from ages ten to thirteen. I was discouraged by this because mathematics was my weakest subject at the time. My father, Edward Walker III, and grandfather took an interest in my love for aviation and encouraged me to continue to pursue my passions.

My father did whatever it took to cultivate my interest in the sciences and aviation. He was once a U.S. Air Force aircraft mechanic for the F-4 Phantom fighter jet and later completed his education at the University of Miami, earning a master's in business administration. My mother, Juanita Walker, was a police officer for the city of Miami police department and an entrepreneur, establishing Sheyes of Miami learning centers. My mother was instrumental in cultivating my interest in aerospace. She prayed over me and my siblings (I'm the oldest of three) and instilled us with words of affirmation of future greatness and encouragement.

In high school, my first introduction to cosmology was a book titled *Black Holes and Warped Spacetime* by William J. Kaufmann. This gave me the push to tackle mathematics going into college. Needless to say, I struggled. I started out pursuing a bachelor's degree in aeronautical science and later decided to pursue aerospace engineering at Embry-Riddle Aeronautical university. My two semesters at Embry-Riddle ended in disaster. The only thing that I would accomplish there was becoming a member of Alpha Phi Alpha fraternity. However, I pressed on, and after passing my first course in calculus, my relationship and love for mathematics grew.

I continued studying mathematics at Florida Atlantic University (Go, Owls!) and later transferred. Thus, after persevering through my issues and senior level mathematics courses, I graduated from

Florida Memorial University with a bachelor's in mathematics and later earned my master's in instructional technology from Grand Canyon University. My study of advanced physics began in 2011 independently; I was able to fully grasp and derive Einstein's field equations by the year 2014 and later quantum mechanics and quantum field theory. After reading about Mexican physicist Miguel Alcubierre's publication about warp fields in terms of general relativity, I was inspired to conduct research and publish.

The path to publishing was difficult and riddled with rejection. Once more, I persevered, and in 2016 I published my first paper, titled "Gravitational Space-Time Curves Generation via Accelerated Particles" in the *Journal of Modern Physics*. I went on to publish several more papers in various journals (including the cosmological papers introducing my inverse gravity concept, which this book is based on) and wrote preprints which are posted on *ResearchGate*. My publications attracted a moderate number of citations from other researchers (at the time of this writing), and I have reviewed/refereed for many journals including publishers such as Springer Nature Physics and the American Institute of Physics (AIP).

The Inverse Gravity Concept

My interest in cosmology had been rekindled after two publications in physics. I was helping my sister with her move from North Carolina to South Florida after she graduated from law school when I wondered about cosmological expansion and current theories that attempted to explain the unknowns. As I walked back and forth toting my sister's possessions, I asked myself, "What if everyone is overthinking the mathematical structure of cosmological expansion?" Moreover, I asked, "What if the root to cosmological expansion is simply the inverse of classical gravity on both macroscopic and quantum level scales?" So I began jotting down and formulating equations that correlate to the simplistic idea and description of the cosmos and applied them to established concepts and equations of cosmology. Therefore, the inverse gravity cosmological concept is an attempt to explain the inflation of the nascent universe in the early epochs of its development to its current expansion. In synopsis of the concept, a sum of classical Newtonian gravitational force and a parameterized inverse Newtonian gravitational force term (e.g., r^2/GMm) is the mathematical basis of the gravity versus expansion model of the universe which will be expressed in detail in chapter 2.

As with all my publications, including my first book titled *Warp Field Mechanics and the Possibility of FTLT (revised): A Description by the Velocity-Based Gravitation by Accelerated Particles (VBGAP) Metric*, I used a particular strategy. This strategy is, as I mentioned, presenting established concepts in physics and their corresponding mathematical descriptions and then adding and/or comparing the new idea and mathematics to it. Hence, I spend a copious amount of text explaining established equations, mathematical expressions, and concepts prior to incorporating the theoretical inverse gravity concept. Implementing this process, the theoretical basis of the inverse gravity cosmological concept is applied to the notions of dark energy, weakly interacting massive particles (WIMPs), the theoretical graviton, Friedman-Walker-Robinson cosmology, and linearized gravity. Thus, this book expounds on the content of three of my published papers:

1. The Inverse Gravity Inflationary Theory of Cosmology

2. Cosmology: The Theoretical Possibility of Inverse Gravity as a Cause of Cosmological Inflation in an Isotropic and Homogenous Universe and Its Relationship to Weakly Interacting Massive Particles.

3. The Relationship between the Cosmological Inverse Gravity Assertion and the Cosmological Constant Including an Alternative Possibility of the Graviton.

Therefore, a partial goal of this book is to improve upon the deficiencies of the seminal papers and to expound upon their strengths. Hence, I want to express the failures in the "technical correctness" of the theoretical concept as it developed. The ultimate goal of this book, as with its predecessors, is to show the incorporation of the inverse gravity concept into established notions and mathematics in cosmology that have some basis and affirmation in cosmological observation.

The first paper published, which introduced the inverse gravity cosmology concept, was titled "The Inverse Gravity Inflationary Theory of Cosmology" or #1 as previously mentioned. Despite its strengths, this paper is riddled with mathematical and conceptual inaccuracies and published in a journal with a poor reputation. Moreover, the paper focused on cosmological expansion over the inflationary epochs (which contradicted the premise expressed in the paper) of the nascent universe, where cosmological expansion transpires beyond the accelerated early epochs of cosmological inflation. However, the paper featured many correct concepts in cosmology and in the mathematics of general relativity, which ultimately laid the foundation for the growth of the inverse gravity cosmological concept. The first paper published on the inverse gravity cosmological concept was essential to my growth and understanding and constituted great practice with cosmology as well as physics in general.

The James Webb Space Telescope

Image credit: https://webb.nasa.gov/

I would be remiss to publish a book on the topic of cosmology and fail to acknowledge the James Webb space telescope (JWST). The James Webb was launched on December 25, 2021, from

French Guiana on an Ariane 5 rocket. At a cost of ten billion dollars, the James Webb telescope was launched with a mission to observe the nascent universe and the formation of exoplanets. After thirty days in space, the JWST traveled nearly a million miles to reach Lagrange point 2, which is a gravitationally stable position in space.

The James Webb telescope focuses on four aspects of the universe: the first light of the universe, the assembly of galaxies in the nascent universe, the creation of stars and protoplanets, and planetary systems. The observation of galaxies gives useful data on the organization of matter on cosmological scales as well as the evolution of the structure of the universe. As an example of observations and data obtained by the James Webb telescope, there can be observations of the shift in the shapes of galaxies such as the evolution of spiral and elliptical galaxies over billions of years. Hence, using data acquired from the JWST allows scientists to make comparisons in data that convey the development of the universe (Dobrijevic & Howell, 2023).

At the time of this writing, the James Webb telescope has observed galaxies unexpectedly old or too far along in their development within the early epochs of the universe, contradicting predictions by the cosmological models of the time (these galaxies are known as "universe breakers"). This challenged current cosmological models and theoretical notions. However, the expansion of the universe is still verified, at least at the time of this writing. Therefore, the inverse gravity cosmological concept presented in this book is still a theoretical candidate explanation for the phenomenon of cosmological expansion in the age of the James Webb telescope.

Chapter 2

The Inverse Gravity Concept

In chapter 2, I present the concept and mathematics of the inverse gravity concept first published in my paper titled: "The Inverse Gravity Inflationary Theory of Cosmology," mentioned in chapter 1. As expressed in chapter 1, I delight in the opportunity to present the inverse gravity cosmology concept and improvements to my proposed (and thus, hypothesized) theoretical concept of the cosmos. Thus, we begin with the classical expression of Newtonian gravity with gravitational force $F_g(r)$ as shown below (Young & Freedman, 2004).

$$F_g(r) = \frac{Gm_1 m_2}{r^2} \qquad (2.0)$$

Where G is the gravitational constant, masses m_1 and m_2 are the gravitational mass values, and r is the distance between masses m_1 and m_2. Therefore, Newtonian gravity or the classical expression of the force of gravity $F_g(r)$ constitutes gravity on a local level. At this juncture, I propose a gravitational inverse that accounts for cosmological expansion on a cosmological level which will be denoted as $F_g'(r)$. Hence, the parameterized inverse gravity term $F_g'(r)$ takes on a value such that (Walker, 2017):

$$F_g'(r) = \left[\frac{1}{r_0^2}\right]\left[\frac{r^2}{Gm_1 m_2}\right] \qquad (2.01)$$

The theoretical parameterized inverse gravity term $F_g'(r)$ is expressed in junction with the classical expression of Newton's equation of force $F_g(r)$; hence, we have what will be referred to as the Newtonian correction. Thus, the proposed gravitational term $F_T(r)$ constitutes the totality of gravity on a cosmological level and is the difference between the inverse gravity term $F_g'(r)$ and Newton's equation of force $F_g(r)$. Cosmological force $F_T(r)$ and thus, the Newtonian correction as presented in the seminal papers, is expressed such that (Walker, 2017):

$$F_T(r) = F_g'(r) - F_g(r) = \left[\frac{1}{r_0^2}\right]\left[\frac{r^2}{Gm_1\,m_2}\right] - \frac{Gm_1\,m_2}{r^2} \tag{2.02}$$

where the inverse term describes the expansion of the universe and possesses a parameter coefficient $[1/r_0^2]$ that adheres to the inequality of (Walker, 2017):

$$1 > \left[\frac{1}{r_0^2}\right] \tag{2.03}$$

Thus, for parameter r_0 such that $r_0 \in R$ (i.e. the set of real numbers), the Newtonian correction $F_T(r)$ has a condition of (Walker, 2017):

1. $For\ r > r_0\ ;\ + F_T(r)$ (2.04)
2. $For\ r < r_0\ ;\ - F_T(r)$ (2.05)

Hence, for values of radial distance r relative to parameter r_0, the Newtonian correction $F_T(r_0)$ increases in value (i.e., $+ F_T(r)$) or decreases in value (i.e., $- F_T(r)$). Figure 1 below is taken from my original inverse gravity papers and shows the relationship between radial distance r and parameter r_0 and expansion gravitational force $+ F_T(r)$ and contraction gravitational force $- F_T(r)$ of Eqs. 2.04–2.05 (Walker, 2017).

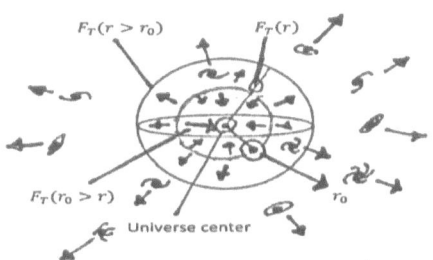

Figure 1

Therefore, we can now define the value of parameter r_0. The value of parameter r_0 is obtained when the inverse gravity term $F_g'(r_0)$ is set equal to the classical gravity term $F_g(r_0)$ which gives a Newtonian correction value of zero or $F_T(r_0) = 0$ where distance r is set equal to parameter distance r_0 ($r = r_0$) (Walker, 2017).

$$F_T(r_0) = F_g'(r_0) - F_g(r_0) = 0 \qquad (2.06)$$

This can alternatively be expressed such that (Walker, 2017):

$$F_T(r_0) = \left[\frac{1}{r_0^2}\right]\left[\frac{(r_0)^2}{Gm_1\,m_2}\right] - \frac{Gm_1\,m_2}{(r_0)^2} = 0 \qquad (2.07)$$

and therefore,

$$\left[\frac{1}{r_0^2}\right]\left[\frac{(r_0)^2}{Gm_1\,m_2}\right] - \frac{Gm_1\,m_2}{(r_0)^2} = 0 \qquad (2.08)$$

which gives us a parameter r_0 value of (Walker, 2017):

$$r_0 = Gm_1\,m_2 \qquad (2.09)$$

The original papers conveyed that the parameter distance r_0 resides at a universal center. This was incorrect; the parameter distance r_0 has a midpoint at the center of the universe. Thus, although the parameter distance r_0 can have a midpoint at the center of the universe, it can be applied to various local positions (this was not specified in the original papers). The product of the gravitational constant G and masses m_1 and m_2 (Gm_1m_2) is approximately equal to the integral of (Walker, 2017):

$$Gm_1\,m_2 \approx G\left[\int_0^{m_u}\int_0^{\pi}\int_0^{\pi} m(m_u'm_u')\,dm d\theta d\phi\right] \qquad (2.10)$$

Eliminating the gravitational constant G from the expression above gives (Walker, 2017):

$$m_1\,m_2 \approx \left[\int_0^{m_u}\int_0^{\pi}\int_0^{\pi} m(m_u'm_u')\,dm d\theta d\phi\right] \qquad (2.11)$$

Expressing the integral above as a product of masses m_1 and m_2 seems to be taboo because the expression describes not merely the gravitational interaction between masses m_1 and m_2 but the gravitational interaction of matter evenly distributed over a spherical surface of a given thickness separated by distance r_0. Figure 2 below shows the distribution of matter/mass over a spherical universe where the variations in mass (dm) transpire as a continuous sum over angles θ and ϕ as rotation about the midpoint of distances r and r_0 over 180^0 or π on the horizontal and vertical axis. As shown in Figure 2 below, each rotation of π is half of the spherical region on both sides of the midpoint of distances r and r_0 where both halves equal 2π or $360^{0,}$ which constitutes the entire sphere or in this case, universe.

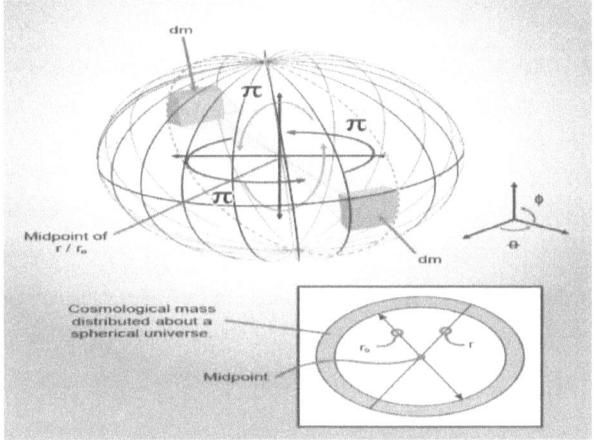

Figure 2

Therefore, I introduce a new notation $M_A'M_B'$ (not done in the seminal papers), which signifies the gravitational interaction between masses m_1 and m_2 as an integral describing the uniform distribution of mass about the spherical region of an isotropic universe. The notation $M_A'M_B'$ distinguishes itself from mass product m_1m_2 in that it represents the continuous sums (or integral) of mass distributed over a spherical surface separated by parameter distance r_0, thus, notation $M_A'M_B'$ is minimally substituted for mass product m_1m_2 shown below.

13

$$M'_A M'_B \rightarrow m_1 m_2 \tag{2.12}$$

Hence, the spherical mass integral can be expressed in terms of mass integral $M'_A M'_B$ such that:

$$M'_A M'_B = \left[\int_0^{m_u} \int_0^{\pi} \int_0^{\pi} m(m'_u m'_u) \, dm \, d\theta \, d\phi \right] \tag{2.13}$$

This will cause a slight nuance in the inverse gravity cosmology expressions as compared to the original papers from this point on. Hence, the Newtonian correction is restated as:

$$F_T(r) = \left[\frac{1}{r_0^2} \right] \left[\frac{r^2}{GM'_A M'_B} \right] - \frac{GM'_A M'_B}{r^2} \tag{2.14}$$

where the spherical distribution of cosmological mass is diametrically separated by a distance r relative to parameter distance r_0. Thus, the value of parameter r_0 is given such that (Walker, 2017):

$$r_0 = GM'_A M'_B \tag{2.15}$$

where, once more, the notation $M'_A M'_B$ is the exact value of the triple integral describing the gravitational interaction of mass uniformly distributed over the surface area of a sphere with a diameter of r_0. Calculations of r_0 will constitute approximations based on astronomical observations; thus we have the expression of (Walker, 2017):

$$r_0 \approx G \left[\int_0^{m_u} \int_0^{\pi} \int_0^{\pi} m(m'_u m'_u) \, dm \, d\theta \, d\phi \right] \tag{2.16}$$

However, from a mathematical standpoint, the value of r_0 will be exact, giving an equal expression such that (Walker, 2017):

$$r_0 = G \left[\int_0^{m_u} \int_0^{\pi} \int_0^{\pi} m(m'_u m'_u) \, dm \, d\theta \, d\phi \right] \tag{2.17}$$

where m_u is the mass value of all the mass in the universe and ρ_c is the critical density of the universe which is 27% of the total density of the universe. Relativistic energy E below can be expressed as (Young & Freedman, 2004):

$$E = m_u c^2 = (.27)\rho_c c^2 \tag{2.18}$$

where c is the velocity of light. The value of the critical density ρ_c is given such that (Walker, 2017):

$$\rho_c = \frac{3H_0^2}{8\pi G} \tag{2.19}$$

Where H_0 is Hubble's constant and G is the gravitational constant, mass m_u assumes a value of (Walker, 2017):

$$m_u = (.27)\rho_c \equiv (.27)\frac{3H_0^2}{8\pi G} \tag{2.20}$$

Therefore, the upper limit value of cosmological mass m_u to the integral in respect to mass m is defined. Now consider the triple integral of (Walker, 2017):

$$M_A' M_B' = \left[\int_0^{m_u} \int_0^{\pi} \int_0^{\pi} m(m_u' m_u') \, dm d\theta d\phi \right] \tag{2.21}$$

where variant mass m_u' in the equation above takes on a value such that (Walker, 2017):

$$m_u' = [(m\cos\theta\sin\phi)^2 + (m\sin\theta\sin\phi)^2 + (m\cos\phi)^2]^{1/2} \tag{2.22}$$

Variable mass m_u' above is a function of spherical coordinates at θ and ϕ, and mass m is a variation in mass in reference to carrying out the operation of integration. The spherical distribution of mass $M_A' M_B'$ on a cosmological scale can alternatively be expressed such that:

$$M_A' M_B' = \int_0^{m_u} \int_0^{\pi} \int_0^{\pi} m^3 [(\cos\theta\sin\phi)^2 + (\sin\theta\sin\phi)^2 + (\cos\phi)^2] \, dm d\theta d\phi \tag{2.23}$$

15

Recall that parameter distance r_0 can be expressed such that:

$$r_0 = GM_A'M_B' \tag{2.24}$$

The parameter distance r_0 can alternatively be expressed such that (Walker, 2017):

$$r_0 = G \int_0^{m_u} \int_0^{\pi} \int_0^{\pi} m^3 [(cos\theta sin\phi)^2 + (sin\theta sin\phi)^2 + (cos\phi)^2] \, dm d\theta d\phi \tag{2.25}$$

where carrying out the integration will be left as an exercise. Conventionally, gravitational potential energy, which will be denoted $U_g(r)$, is the integral of Newtonian gravitation $F_g(r)$ with respect to distance r. Thus, gravitational potential energy can be expressed as (Young & Freedman, 2004):

$$U_g(r) = \int F_g(r)dr \equiv \frac{Gm_1 m_2}{r} \tag{2.26}$$

where, once more, classical Newtonian gravity $F_g(r)$ is given such that:

$$F_g(r) = \frac{Gm_1 m_2}{r^2} \tag{2.27}$$

Hence, gravitational potential energy, denoted $U_T(r)$ in terms of the Newtonian correction $F_T(r)$, which incorporates the inverse gravity term, can be expressed as an integral with respect to distance r such that (Walker, 2017):

$$U_T(r) = \int F_T(r)dr = \int \left[\left[\frac{1}{r_0^2}\right]\left[\frac{r^2}{GM_A'M_B'}\right] - \frac{GM_A'M_B'}{r^2} \right] dr \tag{2.28}$$

With the definition of gravitational potential energy in terms of the inverse gravity concept, we are prepared to extend the theoretical concept to cosmology which will be presented in the next chapter.

Chapter 3

The Inverse Gravity Cosmology

In chapter 3, we define cosmology in terms of the inverse gravity concept. I will begin by applying the inverse gravity concept to an isotropic and homogeneous model in terms of a Friedman-Lemaitre-Walker-Robertson universe. Thus, after evaluating the integral of gravitational potential energy in terms of the inverse gravity concept, the expression of gravitational potential energy in terms of the inverse gravity concept expressed in the original papers is shown below (Walker, 2017).

$$U_T(r) = \left[\frac{1}{r_0^2}\right]\left[\frac{r^3}{3GM_A'M_B'}\right] + \frac{GM_A'M_B'}{r} \tag{3.0}$$

After further thought, I found that the expression of gravitational potential energy in terms of the inverse gravity concept is fundamentally incorrect from a mathematical standpoint. The problem with this is that inverse gravity cosmology potential energy $U_T(r)$ is expressed as an indefinite integral, or an integral with no upper or lower limits of integration. This means that there is a constant of integration, in line with conventional calculus. We include the constant of integration C such that:

$$\int F_T(r)dr = \left[\frac{1}{r_0^2}\right]\left[\frac{r^3}{3GM_A'M_B'}\right] + \frac{GM_A'M_B'}{r} + C \tag{3.01}$$

This implies that gravitational potential energy in terms of the inverse gravity assertion $U_T(r)$ can be expressed including the constant of integration C as shown below.

$$U_T(r) = \left[\frac{1}{r_0^2}\right]\left[\frac{r^3}{3GM_A'M_B'}\right] + \frac{GM_A'M_B'}{r} + C \tag{3.02}$$

To distinguish the notation, the constant of integration C is set equal to or substituted with the notation W_r, as expressed below.

$$C = W_r \tag{3.03}$$

Thus, inverse gravity potential energy $U_T(r)$ can be expressed in terms of constant W_r such that:

$$U_T(r) = \left[\frac{1}{r_0^2}\right]\left[\frac{r^3}{3GM_A'M_B'}\right] + \frac{GM_A'M_B'}{r} + W_r \tag{3.04}$$

Where constant W_r assumes a value such that:

$$W_r = U_T(r) - \left[\frac{1}{r_0^2}\right]\left[\frac{r^3}{3GM_A'M_B'}\right] - \frac{GM_A'M_B'}{r} \tag{3.05}$$

Constant W_r assumes any real number value ($W_r \in R$). The addition of constant W_r to the expression of inverse gravity potential energy $U_T(r)$ is another beautiful opportunity afforded me by writing this book. I must mention that constant W_r can take on values in other terms, as will be shown shortly. Thus, with the establishment of gravitational potential energy in terms of the inverse gravity concept, I now move on to gravitational redshift in the goal of defining the inverse gravity cosmology concept in terms of an isotropic and homogeneous model of a Friedman-Lemaitre-Walker-Robertson universe. The gravitational red shift equation is expressed such that (Walker, 2017):

$$z = \frac{U_g(r) - E_0}{E_0} = \frac{\left(\frac{Gm_1 m_2}{r}\right) - E_0}{E_0} \tag{3.06}$$

When researching and writing the original papers, it was incumbent on me to express the difference between the Schwarzschild expression of gravitational redshift and gravitational redshift z. Thus, as expressed in the seminal papers, I want to distinguish the mathematical expression of gravitational redshift localized to a spherical body of mass M where radius values of r_R and r_E are given by the equation of (Walker, 2017):

$$\frac{\lambda_R}{\lambda_E} - 1 = \sqrt{\frac{1 - 2GM/r_R}{1 - 2GM/r_E}} - 1 \tag{3.07}$$

where λ_R and λ_E are wavelengths that correspond to gravitationally influenced light propagating in the vicinity of a spherical massive body. Conversely, red shift z describes the stretching of a photon (and thus its wavelength) due to the expansion of spacetime on a cosmological and nonlocal level. Red shift z in terms of wavelength can be expressed such that (Walker, 2017):

$$z = \frac{U_g(r) - E_0}{E_0} = \frac{\left(\frac{hc}{\lambda_g}\right) - E_0}{E_0} \equiv \frac{\lambda_0}{\lambda_g} - 1 \qquad (3.08)$$

where initial energy E_0 takes on a value such that:

$$E_0 = \frac{hc}{\lambda_0} \qquad (3.09)$$

where h is Planck's constant, c is the velocity of light, and λ_0 the wavelength at the time of emission. Once more, gravitational potential energy $U_g(r)$ (where $U_g(r) = U_T(r)$) in terms of the inverse gravity concept is given by:

$$U_g(r) = \left[\frac{1}{r_0^2}\right]\left[\frac{r^3}{3GM_A'M_B'}\right] + \frac{GM_A'M_B'}{r} + W_r \qquad (3.10)$$

This gives redshift z in terms of the inverse gravity cosmology concept $U_g(r)$ such that (Walker, 2017):

$$z = \frac{U_g(r) - E_0}{E_0} \qquad (3.11)$$

which can be expressed such that (Walker, 2017):

$$z = \frac{1}{E_0}\left(\left[\frac{1}{r_0^2}\right]\left[\frac{r^3}{3GM_A'M_B'}\right] + \frac{GM_A'M_B'}{r} + W_r\right) - \frac{E_0}{E_0} \qquad (3.12)$$

This reduces to (Walker, 2017):

$$z = \frac{1}{E_0}\left(\left[\frac{1}{r_0^2}\right]\left[\frac{r^3}{3GM_A'M_B'}\right] + \frac{GM_A'M_B'}{r} + W_r\right) - 1 \qquad (3.13)$$

Once more, the energy of a photon is expressed such that (Walker, 2017):

$$E_{ph} = \frac{hc}{\lambda_g} \qquad (3.14)$$

Therefore, a photon propagating through expanding and/or contracting space as described by the inverse gravity cosmological concept can be expressed as the equivalence of (Walker, 2017):

$$\frac{hc}{\lambda_g} = U_T(r) \qquad (3.15)$$

Where $U_T(r)$ is gravitational potential energy in terms of the inverse gravity cosmological concept, this can alternatively be expressed such that (Walker, 2017):

$$\frac{hc}{\lambda_g} = \left[\frac{1}{r_0^2}\right]\left[\frac{r^3}{3GM_A'M_B'}\right] + \frac{GM_A'M_B'}{r} + W_r \qquad (3.16)$$

Hence, the wavelength corresponding to the inverse gravity cosmology concept denoted λ_g takes on a value such that (Walker, 2017):

$$\lambda_g = hc\left[\left[\frac{1}{r_0^2}\right]\left[\frac{r^3}{3GM_A'M_B'}\right] + \frac{GM_A'M_B'}{r} + W_r\right]^{-1} \qquad (3.17)$$

The inverse gravity concept wavelength λ_g will find further use in chapter 4, defining the cosmic microwave background in terms of the inverse gravity cosmological concept. Thus, red shift value z is related to the scale factor of the past (or the time of emission) denoted $a(t_{em})$ and the present scale factor denoted a_0 (which is typically set equal to one, $a_0 = 1$) such that (Walker, 2017):

$$1 + z = \frac{a_0}{a(t_{em})} \qquad (3.18)$$

Substituting the inverse gravity value of red shift value z given by:

$$z = \frac{1}{E_0}\left(\left[\frac{1}{r_0^2}\right]\left[\frac{r^3}{3GM_A'M_B'}\right] + \frac{GM_A'M_B'}{r} + W_r\right) - 1 \qquad (3.19)$$

into the scale factor equation (Eq.3.18) gives (Walker, 2017):

$$1 + \left(\left(\frac{1}{E_0} \left[\frac{1}{r_0^2} \right] \left[\frac{r^3}{3GM_A'M_B'} \right] + \frac{GM_A'M_B'}{r} + W_r \right) - 1 \right) = \frac{a_0}{a(t_{em})} \qquad (3.20)$$

which reduces to (Walker, 2017):

$$\left(\frac{1}{E_0} \left[\frac{1}{r_0^2} \right] \left[\frac{r^3}{3GM_A'M_B'} \right] + \frac{GM_A'M_B'}{r} + W_r \right) = \frac{a_0}{a(t_{em})} \qquad (3.21)$$

Hence, as expressed earlier, constant W_r can take on other values. Therefore, the value of constant W_r can alternatively be expressed such that:

$$W_r = \frac{a_0 E_0}{a(t_{em})} - \left[\frac{1}{r_0^2} \right] \left[\frac{r^3}{3GM_A'M_B'} \right] - \frac{GM_A'M_B'}{r} \qquad (3.22)$$

Thus, the scale factor $a(t_{em})$ at the time of radiation emission in terms of the inverse gravity cosmological concept is expressed such that (Walker, 2017):

$$a(t_{em}) = a_0 \left[\frac{1}{E_0} \left[\left[\frac{1}{r_0^2} \right] \left[\frac{r^3}{3GM_A'M_B'} \right] + \frac{GM_A'M_B'}{r} + W_r \right] \right]^{-1} \qquad (3.23)$$

Defining the scale factor $a(t_{em})$ in terms of the corrected inverse gravity cosmology concept presents an opportunity to make a correlation to the cosmological energy density value Ω_0. The cosmological energy density value Ω_0 defines a mathematical relationship to the early epochs of the universe. Hence, the total density parameter of the universe Ω_0 can be expressed such that (Profound physics.com, 2023):

$$\Omega_0 = \Omega_R + \Omega_M + \Omega_\Lambda \qquad (3.24)$$

where Ω_R is the density parameter of radiation, Ω_M is the density parameter of matter, and Ω_Λ is the density parameter of cosmological expansion or dark energy. The density parameters Ω_R, Ω_M, and Ω_Λ have values expressed such that (Profound physics.com, 2023):

$$\Omega_R = \frac{\rho_R}{\rho_c} \qquad \Omega_M = \frac{\rho_m}{\rho_c} \qquad \Omega_\Lambda = \frac{\Lambda}{3H_0^2} \qquad (3.25)$$

where ρ_c (as previously expressed) is the critical density of the universe, ρ_R and ρ_M are radiation and matter densities respectively, Λ denotes the cosmological constant, and H_0 is Hubble's constant. The mathematical expression for past and future epochs in terms of density parameter correlating to a Friedman-Lemaitre-Walker-Robertson model is given such that (Profound physics.com, 2023):

$$\Omega = \Omega_R x^4 + \Omega_M x^3 + \Omega_\Lambda \qquad (3.26)$$

Where variable x relate to the scale factor such that (Profound physics.com, 2023):

$$x = \frac{a_0}{a(t_{em})} \qquad (3.27)$$

This implies that if:

$$\frac{a_0}{a(t_{em})} = a_0 \left[\frac{1}{E_0} \left[\left[\frac{1}{r_0^2} \right] \left[\frac{r^3}{3GM_A'M_B'} \right] + \frac{GM_A'M_B'}{r} + W_r \right] \right]^{-1} \qquad (3.28)$$

Then variable x in terms of the inverse gravity cosmological concept is given such that:

$$x = a_0 \left[\frac{1}{E_0} \left[\left[\frac{1}{r_0^2} \right] \left[\frac{r^3}{3GM_A'M_B'} \right] + \frac{GM_A'M_B'}{r} + W_r \right] \right]^{-1} \qquad (3.29)$$

Therefore, the density parameter Ω for a Friedman-Lemaitre-Walker-Robertson model in terms of the inverse gravity cosmological concept can be expressed such that:

$$\Omega = \Omega_R \left[a_0 \left[\frac{1}{E_0} \left[\left[\frac{1}{r_0^2} \right] \left[\frac{r^3}{3GM_A'M_B'} \right] + \frac{GM_A'M_B'}{r} + W_r \right] \right]^{-1} \right]^4 + \Omega_M \left[a_0 \left[\frac{1}{E_0} \left[\left[\frac{1}{r_0^2} \right] \left[\frac{r^3}{3GM_A'M_B'} \right] + \frac{GM_A'M_B'}{r} + W_r \right] \right]^{-1} \right]^3 + \Omega_\Lambda$$

$$\qquad (3.30)$$

In transitioning to another facet of the Friedman-Lemaitre-Walker-Robertson model in terms of the inverse gravity cosmology concept, the Friedman-Lemaitre-Walker-Robertson metric is shown below (Walker, 2017).

$$d\Sigma^2 = -dt^2 + a^2(t)\left[\frac{dr^2}{1-kr^2} + r^2 d\theta^2 + r^2 sin^2\theta d\phi^2\right] \tag{3.31}$$

Where k is equal to +1 ($k = +1$) for the positive spherical curvature of space-time for the description of expanding space-time, the scale factor $a(t)$ above is set equal to the scale factor $a(t_{em})$ such that:

$$a(t) = a(t_{em}) \tag{3.32}$$

Where $t = t_{em}$, the Friedman-Lemaitre-Walker-Robertson metric can be expressed such that (Walker, 2017):

$$d\Sigma^2 = -dt^2 + a^2(t_{em})\left[\frac{dr^2}{1-kr^2} + r^2 d\theta^2 + r^2 sin^2\theta d\phi^2\right] \tag{3.33}$$

where once more, the scale factor $a(t_{em})$ within the Friedman-Lemaitre-Walker-Robertson metric has a value in terms of the inverse gravity cosmology concept such that (Walker, 2017):

$$a(t) = a(t_{em}) = a_0\left[\frac{1}{E_0}\left[\left[\frac{1}{r_0^2}\right]\left[\frac{r^3}{3GM_A'M_B'}\right] + \frac{GM_A'M_B'}{r} + W_r\right]\right]^{-1} \tag{3.34}$$

Hence, the Friedman-Lemaitre-Walker-Robertson metric can be expressed in terms of the inverse gravity cosmological concept as shown below (Walker, 2017).

$$d\Sigma^2 = -dt^2 + \left[a_0\left[\frac{1}{E_0}\left[\left[\frac{1}{r_0^2}\right]\left[\frac{r^3}{3GM_A'M_B'}\right] + \frac{GM_A'M_B'}{r} + W_r\right]\right]^{-1}\right]^2\left[\frac{dr^2}{1-kr^2} + r^2 d\theta^2 + r^2 sin^2\theta d\phi^2\right]$$

$$\tag{3.35}$$

The cosmological distance r expressed in subsequent equations relates to the three spatial coordinates x, y, z and one temporal component t. Thus, the changes in the components $(x, y, z,$ and $t)$ relate to each other according to the Pythagorean equation of (Wald, 1984):

$$\Delta r^2 = -\Delta t^2 + \Delta x^2 + \Delta y^2 + \Delta z^2 \tag{3.36}$$

where the metric has a signature of $-+++$, and the terms $\Delta t, \Delta x, \Delta y,$ and Δz are the differences between final and initial values such that:

$$\Delta t = t_f - t_i \qquad \Delta x = x_f - x_i \qquad \Delta y = y_f - y_i \qquad \Delta z = z_f - z_i \tag{3.37}$$

My original papers were a bit myopic in reference to the equations above and were missing the detail expressed in the equations above. Hence, I am grateful for this opportunity in this book. Thus, the change in radial distance Δr can be expressed in terms of final and initial values such that:

$$\Delta r = r_f - r_i \tag{3.38}$$

The inverse gravity concept assumes an origin point. Hence, all values of $\Delta t, \Delta x, \Delta y,$ and Δz have initial values of zero as shown below.

$$r_f = r \quad r_i = 0 \tag{3.39}$$

$$t_f = t \quad t_i = 0 \tag{3.40}$$

$$x_f = x \quad x_i = 0 \tag{3.41}$$

$$y_f = y \quad y_i = 0 \tag{3.42}$$

$$z_f = z \quad z_i = 0 \tag{3.43}$$

where the final values of Δr, Δt, Δx, Δy, and Δz are arbitrary values of r, x, y, and z as shown by the equations above. Hence, values of Δr, Δt, Δx, Δy, and Δz are expressed such that:

$$\Delta t = t_f - 0 \qquad \Delta x = x_f - 0 \qquad \Delta y = y_f - 0 \qquad \Delta z = z_f - 0 \qquad (3.44)$$

Hence, the distance formula becomes:

$$r^2 = -t^2 + x^2 + y^2 + z^2 \qquad (3.45)$$

where the value of radial distance r becomes:

$$r = [-t^2 + x^2 + y^2 + z^2\ ]^{\frac{1}{2}} \qquad (3.46)$$

for the condition of:

$$t \leq x^2 + y^2 + z^2 \qquad (3.47)$$

where if violated, distance r will fall into the undefined realm of imaginary numbers. The value of r is substituted into scale factor $a(t_{em})$ in terms of the inverse gravity cosmological concept $(a(t_{em}) \rightarrow a(t, x, y, z))$ and gives a value such that (Walker, 2017):

$$a(t_{em}) = a(t, x, y, z) = a_0 \left[\frac{1}{E_0}\left[\left[\frac{1}{r_0^2}\right]\left[\frac{(-t^2+x^2+y^2+z^2)^{3/2}}{3GM_A'M_B'}\right] + \frac{GM_A'M_B'}{[-t^2+x^2+y^2+z^2]^{1/2}} + W_r\right]\right]^{-1}$$

$$(3.48)$$

Thus, recall that scale factor $a(t, x, y, z)$ is simply the scale factor $a(t_{em})$ in respect to the time coordinate t where $t = t_{em}$. Hence, the time derivative of scale factor $a(t, x, y, z)$ is denoted $a(t, \dot{x}, y, z)$ in terms of the Minkowski coordinates and can be expressed such that (Wald, 1984):

$$a(t, \dot{x}, y, z) = \frac{da(t,x,y,z)}{dt} \qquad (3.49)$$

26

This can alternatively be expressed as (Walker, 2017):

$$a(t, \dot{x}, y, z) = \frac{\partial}{\partial t}\left[a_0\left[\frac{1}{E_0}\left[\left[\frac{1}{r_0^2}\right]\left[\frac{(-t^2+x^2+y^2+z^2)^{3/2}}{3M_A'M_B'}\right]\right] + \frac{GM_A'M_B'}{[-t^2+x^2+y^2+z^2]^{1/2}} + W_r\right]\right]^{-1}\right]$$

(3.50)

Applying the chain rule, the time derivative $(da(t, x, y, z))/dt$ of the inverse gravity cosmology scale factor denoted $a(t, \dot{x}, y, z)$ gives a value such that (Walker, 2017):

(3.51)

$$a(t, \dot{x}, y, z) = 2ta_0\left[\frac{1}{E_0}\left[\left[\frac{1}{r_0^2}\right]\left[\frac{(-t^2 + x^2+y^2 + z^2)^{1/2}}{2GM_A'M_B'}\right]\right.\right.$$

$$+ \frac{GM_A'M_B'}{2[-t^2 + x^2+y^2 + z^2]^{3/2}}\right]\left[\frac{1}{E_0}\left[\left[\frac{1}{r_0^2}\right]\left[\frac{(-t^2 + x^2+y^2 + z^2)^{3/2}}{3GM_A'M_B'}\right]\right.\right.$$

$$\left.\left.+ \frac{GM_A'M_B'}{[-t^2 + x^2+y^2 + z^2]^{1/2}} + w_r\right]\right]^{-2}$$

The equation above gives the opportunity to present Hubble's constant in terms of the inverse gravity cosmological concept such that (Wald, 1984):

$$H(t) = \frac{\dot{a}}{a} = \left[\frac{1}{a(t,x,y,z)}\right]\frac{da(t,x,y,z)}{dt}$$

(3.52)

Therefore, Hubble's constant takes on a value in terms of the inverse gravity cosmological concept such that (Walker, 2017):

$$H(t) = 2ta_0 \left[\left[\frac{1}{E_0} \left[\left[\frac{1}{r_0^2} \right] \left[\frac{(-t^2+x^2+y^2+z^2)^{1/2}}{2GM'_A M'_B} \right] + \frac{GM'_A M'_B}{2[-t^2+x^2+y^2+z^2]^{3/2}} \right] \right] \left[\frac{1}{E_0} \left[\left[\frac{1}{r_0^2} \right] \left[\frac{(-t^2+x^2+y^2+z^2)^{3/2}}{3GM'_A M'_B} \right] \right] + \right.$$

$$\left. \frac{GM'_A M'_B}{[-t^2+x^2+y^2+z^2]^{1/2}} + w_r \right]^{-1} \right] \tag{3.53}$$

Thus, in an isotropic and homogeneous universe, the velocity $V(t)$ of cosmological expansion is expressed in reference to Hubble's Law such that (Wald, 1984):

$$V(t) = \frac{Ra(t)}{a(t)} = \left[\frac{R}{a(t)} \right] \frac{da(t)}{dt} \tag{3.54}$$

Where R is the distance from the observer, the velocity of expansion $V(t)$ in terms of the inverse gravity cosmology concept is given such that (Walker, 2017):

$$V(t) = 2tRa_0 \left[\left[\frac{1}{E_0} \left[\left[\frac{1}{r_0^2} \right] \left[\frac{(-t^2+x^2+y^2+z^2)^{1/2}}{2GM'_A M'_B} \right] + \frac{GM'_A M'_B}{2[-t^2+x^2+y^2+z^2]^{3/2}} \right] \right] \left[\frac{1}{E_0} \left[\left[\frac{1}{r_0^2} \right] \left[\frac{(-t^2+x^2+y^2+z^2)^{3/2}}{3GM'_A M'_B} \right] \right] + \frac{GM'_A M'_B}{[-t^2+x^2+y^2+z^2]^{1/2}} + w_r \right]^{-1} \right] \tag{3.55}$$

where the second time derivative of the inverse gravity scale factor is expressed as shown below (Wald, 1984).

$$a(t, \overset{..}{x}, y, z) = \frac{d^2(a(t,x,y,z))}{dt^2} \tag{3.56}$$

In Einstein's field equations pertaining to a geodesic, the value of the symmetric Christoffel symbols Γ^a_{bc} are of the form (Wald, 1984):

$$\Gamma^a_{bc} = \frac{1}{2} \Sigma_d \, g^{ad} \left\{ \frac{\partial g_{cb}}{\partial x^b} + \frac{\partial g_{ca}}{\partial x^c} - \frac{\partial g_{bc}}{\partial x^c} \right\} \tag{3.57}$$

Where g_{cb} are expressions of the four-by-four matrix valued metric tensor and g^{ad} the inverse metric tensor, the scale factors $a(t,x,y,z)$ relate to the symmetric Christoffel symbols Γ^a_{bc} (and thus designating a geodesic) such that (Wald, 1984):

28

$$\Gamma_{xx}^t = \Gamma_{yy}^t = \Gamma_{zz}^t = a(t,x,y,z)a(t,\dot{x},y,z) \tag{3.58}$$

$$\Gamma_{xt}^x = \Gamma_{tx}^x = \Gamma_{ty}^y = \Gamma_{yt}^y = \Gamma_{zt}^z = \Gamma_{tz}^z = a(t,\dot{x},y,z)/a(t,x,y,z) \tag{3.59}$$

where the components of the Ricci tensor are calculated according to the equation of (Wald, 1984):

$$R_{ab} = \sum_c R_{acb}{}^c \tag{3.60}$$

As expressed by Wald, the Ricci tensor above can be expressed in terms of the Christoffel symbols Γ_{bc}^a such that (Wald, 1984):

$$R_{ab} = \sum_c \frac{\partial y}{\partial x^c}\Gamma_{ab}^c - \frac{\partial}{\partial x^a}\left(\sum_c \Gamma_{cb}^c\right) + \sum_{d,c}\left(\Gamma_{ab}^d\Gamma_{dc}^c - \Gamma_{cb}^d\Gamma_{da}^c\right) \tag{3.61}$$

Hence, the expressions of the Ricci tensors R_{tt} and R_{**} are then related to the scale factor $a(t,x,y,z)$ in terms of the inverse gravity cosmological concept by the equations of (Wald, 1984):

$$R_{tt} = -3a(t,\ddot{x},y,z)/a(t,x,y,z) \tag{3.62}$$

$$R_{**} = a(t,x,y,z)^{-2}R_{xx} = \frac{a(t,\ddot{x},y,z)}{a(t,x,y,z)} + 2\frac{(a(t,\dot{x},y,z))^2}{a(t,x,y,z)^2} \tag{3.63}$$

Where $\ddot{a}$ is the second time derivative of the scale factor $a(t,x,y,z)$ in terms of the inverse gravity cosmological concept expressed such that (Wald, 1984):

$$\ddot{a} = \frac{d^2(a(t,x,y,z))}{dt^2} \tag{3.64}$$

As expressed by Wald, and as I expressed in the original papers, the expression of the Ricci tensor R_{xx} (relating to Ricci tensor values of R_{tt} and R_{**}) above in terms of the Christoffel symbols are of the form (Wald, 1984):

$$R_{xx} = \sum_c \frac{\partial y}{\partial x^c} \Gamma_{xx}^c - \frac{\partial}{\partial x^x} \left(\sum_c \Gamma_{cx}^c \right) + \sum_{d,c} \left(\Gamma_{xx}^d \Gamma_{dc}^c - \Gamma_{cx}^d \Gamma_{dx}^c \right) \tag{3.65}$$

Thus, Ricci tensor values R_{tt} and R_{**}, relate to the scalar curvature R such that (Wald, 1984):

$$R = -R_{tt} + 3R_{**} \tag{3.66}$$

Substituting the values of R_{tt} and R_{**} into the equation above gives (Wald, 1984):

$$R = -R_{tt} + 3R_{**} = 6\left(\frac{a(t,\ddot{x},y,z)}{a(t,x,y,z)} + \frac{(a(t,\dot{x},y,z))^2}{a(t,x,y,z)^2} \right) \tag{3.67}$$

which relates the scalar curvature R to the inverse gravity cosmological concept via the scale factor $a(t,x,y,z)$ in terms of the inverse gravity cosmological concept. As expressed by Wald (and in the inverse gravity papers), the values of the Einstein tensor denoted G_{tt} and G_{**} are given such that (Wald, 1984):

$$G_{tt} = \frac{3(a(t,\dot{x},y,z))^2}{a(t,x,y,z)^2} = R_{tt} + \frac{1}{2}R = 8\pi\rho \tag{3.68}$$

$$G_{**} = -2\frac{a(t,\ddot{x},y,z)}{a(t,x,y,z)} - \frac{(a(t,\dot{x},y,z))^2}{a(t,x,y,z)^2} = R_{**} - \frac{1}{2}R = 8\pi P \tag{3.69}$$

Thus, the general evolution of an isotropic and homogeneous model of the universe expressed in terms of the scale factors and thus the inverse gravity cosmological concept is given such that (Wald, 1984):

$$\frac{3(a(t,\dot{x},y,z))^2}{a(t,x,y,z)^2} = 8\pi\rho - \frac{3k}{a(t,x,y,z)^2} \tag{3.70}$$

$$\frac{3a(t,\ddot{x},y,z)}{a(t,x,y,z)} = -4\pi(\rho + 3P) \tag{3.71}$$

Hence, we see the relationship between pressure P corresponding to thermal radiation, the average mass density ρ, and the scale factors $a(t,x,y,z)$ and their corresponding time derivatives ($a(t, \dot{x}, y, z)$ and $a(t, \ddot{x}, y, z)$). Once more, constant k is equal to +1 ($k = +1$) and $r > r_0$ for positive spherical curvature describing the expansion of spacetime in a homogeneous isotropic universe (Wald, 1984).

The Einstein field equations including the cosmological constant Λ in terms of the Friedman-Lemaitre-Walker-Robertson universe for thermal pressure P and the average mass density ρ are shown below (where curvature constant k is equal to +1 ($k = +1$) for $r > r_0$) (Ojeda & Rosu, 2006).

$$\left(\frac{\dot{a}}{a}\right)^2 + \frac{kc^2}{a^2} - \frac{\Lambda c^2}{3} = \frac{8\pi G}{3}\rho \tag{3.72}$$

$$2\left(\frac{\ddot{a}}{a}\right) + \left(\frac{\dot{a}}{a}\right)^2 + \frac{kc^2}{a^2} - \Lambda c^2 = -\frac{8\pi G}{3}P \tag{3.73}$$

Where G is the gravitational constant and c the velocity of light, the values of the cosmological constant Λ can be expressed such that (Walker, 2018):

$$\Lambda = \frac{3}{c^3}\left[\left(\frac{\dot{a}}{a}\right)^2 + \frac{kc^2}{a^2} - \frac{8\pi G}{3}\rho\right] = \frac{1}{c^2}\left[2\left(\frac{\ddot{a}}{a}\right) + \left(\frac{\dot{a}}{a}\right)^2 + \frac{kc^2}{a^2} + \frac{8\pi G}{3}P\right] \tag{3.74}$$

Recall that the scale factor a denoted $a(t, x, y, z)$ in terms of the inverse gravity cosmological concept is expressed such that (Walker, 2018):

$$a(t, x, y, z) = a_0\left[\frac{1}{E_0}\left[\left[\frac{1}{r_0^2}\right]\left[\frac{(-t^2+x^2+y^2+z^2)^{3/2}}{3GM'_A M'_B}\right] + \frac{GM'_A M'_B}{[-t^2+x^2+y^2+z^2]^{1/2}} + W_r\right]\right]^{-1} \tag{3.75}$$

The cosmological constant Λ in terms of the inverse gravity cosmology concept is denoted Λ_{IG}. Hence, the expression of cosmological constant Λ_{IG} in terms of the inverse gravity cosmological concept scale factor $a(t, x, y, z)$ is given such that (Walker, 2018):

$$\Lambda_{IG} = \frac{3}{c^3} \left[\left(\frac{a(t,\dot{x},y,z)}{a(t,x,y,z)} \right)^2 + \frac{kc^2}{\left(a(t,x,y,z) \right)^2} - \frac{8\pi G}{3} \rho \right] \qquad (3.76)$$

$$\Lambda_{IG} = \frac{1}{c^2} \left[2 \left(\frac{a(t,\ddot{x},y,z)}{a(t,x,y,z)} \right) + \left(\frac{a(t,\dot{x},y,z)}{a(t,x,y,z)} \right)^2 + \frac{kc^2}{\left(a(t,x,y,z) \right)^2} + \frac{8\pi G}{3} P \right] \qquad (3.77)$$

Lastly, I will show mathematical application of the inverse gravity cosmology concept to the cosmological equation of state shown below (Profoundphysics.com, 2023).

$$\frac{P}{\rho} = \omega \qquad (3.78)$$

Where P is cosmological pressure and ρ is energy density, the energy density ρ can be expressed such that (Profoundphysics.com, 2023):

$$\rho = \frac{E}{V_c} \qquad (3.79)$$

Where E is the corresponding cosmological energy value and V_c the cosmological density, energy E is set equal to the inverse gravity cosmology concept potential energy $U_T(r)$, as shown below.

$$E = U_T(r) = \left[\frac{1}{r_0^2} \right] \left[\frac{r^3}{3 G M'_A M'_B} \right] + \frac{G M'_A M'_B}{r} + W_r \qquad (3.80)$$

Hence, cosmological density ρ can be expressed in terms of the inverse gravity cosmology concept such that:

$$\rho = \frac{U_T(r)}{V_c} = \frac{\left[\frac{1}{r_0^2} \right] \left[\frac{r^3}{3 G M'_A M'_B} \right] + \frac{G M'_A M'_B}{r} + W_r}{V_c} \qquad (3.81)$$

Alternatively,

$$\rho = \frac{1}{V_c} \left[\left[\frac{1}{r_0^2} \right] \left[\frac{r^3}{3GM_A'M_B'} \right] + \frac{GM_A'M_B'}{r} + W_r \right] \tag{3.82}$$

Thus, pressure P can be expressed such that:

$$P = \omega\rho \tag{3.83}$$

Therefore, substituting the value of density ρ in terms of the inverse gravity cosmology concept into the equation above gives a value of pressure P in terms of the inverse gravity cosmology concept as shown below.

$$P = \frac{\omega}{V_c} \left[\left[\frac{1}{r_0^2} \right] \left[\frac{r^3}{3GM_A'M_B'} \right] + \frac{GM_A'M_B'}{r} + W_r \right] \tag{3.84}$$

The condition of $F_g'(r) > F_g(r)$, where inverse gravity force $F_g'(r)$ is greater than classical Newtonian gravitation $F_g(r)$ on a cosmological scale introduced in chapter 1, implies that dark energy is more prevalent. Thus, the corresponding energy terms of the inverse gravity cosmology concept will satisfy the inequality of:

$$\left[\frac{1}{r_0^2} \right] \left[\frac{r^3}{3GM_A'M_B'} \right] > \frac{GM_A'M_B'}{r} \tag{3.85}$$

Thus, for this condition, constant ω is equal to negative one ($\omega = -1$), and thus, cosmological pressure P adopts a value such that (Profoundphysics.com, 2023):

$$P = -\frac{1}{V_c} \left[\left[\frac{1}{r_0^2} \right] \left[\frac{r^3}{3GM_A'M_B'} \right] + \frac{GM_A'M_B'}{r} + W_r \right] \tag{3.86}$$

Additionally, we see that cosmological pressure P and cosmological density ρ relate to the inverse gravity cosmological concept via energy E and/or the scale factor in terms of the inverse gravity cosmological concept shown below.

$$E = U_T(r) = \left[\frac{1}{r_0^2}\right]\left[\frac{r^3}{3GM_A'M_B'}\right] + \frac{GM_A'M_B'}{r} + W_r \tag{3.87}$$

$$a(t,x,y,z) = a_0 \left[\frac{1}{E_0}\left[\left[\frac{1}{r_0^2}\right]\left[\frac{(-t^2+x^2+y^2+z^2)^{3/2}}{3GM_A'M_B'}\right] + \frac{GM_A'M_B'}{[-t^2+x^2+y^2+z^2]^{1/2}} + W_r\right]\right]^{-1} \tag{3.88}$$

Henceforth, this completes a detailed and holistic mathematical description of cosmological expansion in terms of the inverse gravity cosmological concept for an isotropic and homogeneous model in terms of a Friedman-Lemaitre-Walker-Robertson universe.

Chapter 4

The Inverse Gravity Concept and the Cosmic Microwave Background

The cosmic microwave background or CMB is the "fossil" radiation or simply the remnant of radiation left from the early epochs of the universe; I hesitate to use the term "big bang" in light of observations by the James Webb telescope. CMB radiation was discovered by Arno Penzias and Robert Wilson in the year 1965 using two radio receivers which detected cosmic microwave background radiation. The source of the radiation was determined to have no other emission point than the sky.

Since the beginning of the universe, from a scientific observational standpoint, the universe has expanded and cooled over time. Hence, the CMB has cooled as it expanded and stretched. In the early epochs of the universe, mostly neutrons and charged particles such as protons and electrons dominated the universe. There was a close interaction between electrons and photons, meaning that they were tightly coupled. This implies that light and matter were closely coupled and that light had no ability to travel for long distances in straight lines.

This information has been derived from the observation and measurement of the distribution of cosmic radiation over the universe and gives a general description of the CMB, its contents, and dynamics. In this chapter, I will elucidate the mathematical structure of the CMB. As the CMB indicates the expansion of the universe, the inverse gravity cosmology concept applies. Hence, in this chapter, I will show the incorporation of the inverse gravity cosmology concept to the mathematical structure of the CMB (Winther, 2023).

Figure 3 below shows the cosmic microwave background as viewed by the European Space Agency's Planck observatory.

Image credit: European Space Agency and Planck Collaboration

Figure 3

As the universe expands, the distribution of radiation and spectrum of electromagnetic energy expands and varies with it, which is described by the cosmic microwave radiation power spectrum. Thus, the mathematical structure of the cosmic microwave background is the cosmic microwave radiation power spectrum. Thus, we began a heuristic derivation of the mathematical structure of the CMB power spectrum with the Fourier transform. Hence, the Fourier transform in flat or rectangular space can be expressed such that (Winther, 2023):

$$f(x) = \sum_k [b_k \cos(kx) + ic_k \sin(kx)] = \sum_k a_k e^{ikx} \qquad (4.0)$$

Where b_k and c_k are corresponding constants, and k and x are wavenumber and position respectively. The Fourier coefficient a_k is expressed as the integral in respect to position x such that (Winther, 2023):

$$a_k = \int f(x)e^{-ikx}dx \qquad (4.01)$$

Quantifying the power spectrum of the cosmic microwave background is expressed such that (Winther, 2023):

$$P(k) = |a_k|^2 = a_k^* a_k \tag{4.02}$$

which is the squared modulus of function (or amplitude) a_k, where a_k incorporates the plane wave expression e^{-ikx}, and a_k^* the corresponding complex conjugate ($a_k^* = e^{ikx}$). The power spectrum can be quantified in rectangular coordinates; however, its components are measured in the realm of spherical harmonics. Thus, we move on to spherical harmonics within the expressions of the cosmic microwave background, which are expressed in spherical coordinates θ and φ. Therefore, the plane wave function e^{ikx} of amplitude a_k is minimally substituted for the wave function $Y_{lm}(\theta, \varphi)$ in terms of spherical coordinates θ and φ as shown below (Winther, 2023).

$$e^{ikx} \rightarrow Y_{lm}(\theta, \varphi) \tag{4.03}$$

Where subscript l is the number of waves along a meridian, and subscript m is the number of modes. Here we consider Laplace's equation of the form (Winther, 2023):

$$\nabla^2 \Psi = 0 \tag{4.04}$$

where ∇^2 is the second partial derivative in respect to position, and Ψ represents a wave function or amplitude. Thus, we require Laplace's equation to be expressed in terms of the wave function $Y_{lm}(\theta, \varphi)$ in terms of the spherical coordinates $\boldsymbol{\theta}$ and $\boldsymbol{\varphi}$. Laplace's equation can be presented in spherical coordinates $\boldsymbol{\theta}$ and $\boldsymbol{\varphi}$ pertaining to function $Y_{lm}(\theta, \varphi)$ such that (Winther, 2023):

$$\frac{\Phi_m(\varphi)}{\sin\theta} \frac{d}{d\theta}\left(\sin\theta \frac{d}{d\theta}\right) + \frac{\theta_l^m(\theta)}{\sin^2\theta}\frac{d^2\Phi_m(\theta)}{d\Phi^2} + l(l+1)\theta_l^m(\theta)\Phi_m(\varphi) = 0 \tag{4.05}$$

Where functions $\Phi_m(\varphi)$ and $\theta_l^m(\theta)$ in the Laplace's equation above have values such that (Liboff, 2003):

$$\Phi_m(\varphi) = \frac{e^{im\varphi}}{\sqrt{2\pi}} \qquad (4.06)$$

$$\Theta_l^m(\theta) = \left(\sqrt{\frac{(2l+1)(l-m)!}{2(l+m)!}} \right) P_{lm}(cos\theta) \qquad (4.07)$$

where function $Y_{lm}(\theta, \varphi)$ is a composite or product of both functions $\Phi_m(\varphi)$ and $\Theta_l^m(\theta)$ such that (Liboff, 2003):

$$Y_{lm}(\theta, \varphi) = \Theta_l^m(\theta)\Phi_m(\varphi) \qquad (4.08)$$

This gives to the value of the wave function $Y_{lm}(\theta, \varphi)$ such that (Liboff, 2003):

$$Y_{lm}(\theta, \varphi) = \left(\sqrt{\frac{(2l+1)(l-m)!}{4\pi(l+m)!}} \right) P_{lm}(cos\theta)e^{im\varphi} \qquad (4.09)$$

Hence, we see that there was a separation of variables in the function $Y_{lm}(\theta, \varphi)$ of its terms $\Phi_m(\varphi)$ and $\Theta_l^m(\theta)$ in the application of the Laplacian (Eq. 4.05). For the conditions of l where wave number l is an integer and m is an integer such that $l \geq m \geq \pm l$, the Legendre polynomials $P_{lm}(\mu)$ (i.e. $P_{lm}(cos\theta)$ in Eq. 4.09, where variable $\mu = cos\theta$), are of the form (Liboff, 2003):

$$P_{lm}(\mu) = (-1)^m (1 - \mu)^{m/2} \frac{d^m P_l(\mu)}{d\mu^m} \qquad (4.10)$$

where the series summation solution with the Legendre polynomials of $P_l(\mu)$ in Eq. 4.10 are given such that (Liboff, 2003):

$$P_l(\mu) = \frac{1}{2^l l!} \frac{d^l}{d\mu^l} (\mu^2 - 1)^l \qquad (4.11)$$

Thus, with the definition of the components of function $Y_{lm}(\theta, \varphi)$, any function defined on the spherical region of the universe may be expanded in terms of spherical harmonics such that (Winther, 2023):

$$T(\theta, \varphi) = \sum_{l=0}^{l_{max}} \sum_{m=-l}^{l} a_{lm} Y_{lm}^*(\theta, \varphi) \qquad (4.12)$$

where $Y_{lm}^*(\theta, \varphi)$ is the complex conjugate of function $Y_{lm}(\theta, \varphi)$ and expressed such that:

$$Y_{lm}^*(\theta, \varphi) = \left(\sqrt{\frac{(2l+1)(l-m)!}{4\pi(l+m)!}} \right) P_{lm}(cos\theta) e^{-im\varphi} \qquad (4.13)$$

Hence, the expansion of coefficients a_{lm} in the equation of summation $T(\theta, \varphi)$ is given such that (Winther, 2023):

$$a_{lm} = \int_{4\pi} sin\theta d\theta d\varphi T(\theta, \varphi) Y_{lm}(\theta, \varphi) \qquad (4.14)$$

Thus, terms $T(\theta, \varphi)$ and a_{lm} convey the Fourier transform in terms of spherical coordinates. Hence, introducing the angular power spectrum C_l , the angular power spectrum takes on a value in terms of a_{lm} such that (Winther, 2023):

$$C_l = \frac{2}{2l+1} \sum_{m=-1}^{l} |a_{lm}|^2 \qquad (4.15)$$

Where the angular power spectrum C_l is defined as an average of m for every l, the wavenumber l (e.g. seen in functions a_{lm} and $Y_{lm}(\theta, \varphi)$) relates to wavelength denoted λ with the equation of (Liboff, 2003):

$$l(l + 1) = \lambda \qquad (4.16)$$

This can be expressed as a quadratic equation such that:

$$l^2 + l - \lambda = 0 \qquad (4.17)$$

Applying the quadratic formula to the equation above, one obtains a value of wavenumber l such that:

$$l = \frac{1}{2}\left[-1 \pm [1 + 4\lambda]^{1/2}\right] \tag{4.18}$$

The cosmological distribution of radiation or electromagnetic (or photonic) energy is affected by cosmological expansion. Thus, we can show the correlation of the inverse gravity cosmology concept to the mathematics of the CMB. Radiation propagating across the cosmos composing the cosmological background radiation will, in theory, have an inverse gravity wavelength λ_g as a result of cosmological expansion. Hence, recall the inverse gravity wavelength of λ_g introduced in chapter 3 which takes on a value such that:

$$\lambda_g = hc\left[\left[\frac{1}{r_0^2}\right]\left[\frac{r^3}{3GM_A'M_B'}\right] + \frac{GM_A'M_B'}{r} + W_r\right]^{-1} \tag{4.19}$$

Where constant W_r takes on values such that:

$$W_r = U_T(r) - \left[\frac{1}{r_0^2}\right]\left[\frac{r^3}{3GM_A'M_B'}\right] - \frac{GM_A'M_B'}{r} \tag{4.20}$$

Or

$$W_r = \frac{a_0 E_0}{a(t_{em})} - \left[\frac{1}{r_0^2}\right]\left[\frac{r^3}{3GM_A'M_B'}\right] - \frac{GM_A'M_B'}{r} \tag{4.21}$$

The variations in the CMB correlate to the distance r separating cosmological masses and has a midpoint at the origin (whatever it maybe after the James Webb telescope observations) or central point of reference. Thus, the inverse gravity wavelength λ_g relates to the wavenumber l_g in terms of the inverse gravity concept by the equation of:

$$l_g = \frac{1}{2}\left[-1 \pm \left[1 + 4\lambda_g\right]^{1/2}\right] \tag{4.22}$$

Hence, alternatively,

$$l_g = \frac{1}{2}\left[-1 \pm \left[1 + 4hc\left[\left[\frac{1}{r_0^2}\right]\left[\frac{r^3}{3GM_A'M_B'}\right] + \frac{GM_A'M_B'}{r} + W_r\right]^{-1}\right]^{1/2}\right] \tag{4.23}$$

The wavenumber l_g is an integer value; thus, the value must be rounded up or down to the nearest whole number if the value of wavenumber l_g is not an integer. Therefore, after establishing the value of wavenumber l_g in terms of the inverse gravity concept, we substitute it into the equations of functions $\Phi_m(\varphi)$ and $\Theta_l^m(\theta)$ which gives:

$$\Theta_{l_g}^m(\theta) = \left(\sqrt{\frac{(2l_g+1)(l_g-m)!}{2(l_g+m)!}}\right)P_{l_g m}(cos\theta) \tag{4.24}$$

This gives to the value of the function $Y_{l_g m}(\theta, \varphi)$ such that:

$$Y_{l_g m}(\theta, \varphi) = \left(\sqrt{\frac{(2l_g+1)(l_g-m)!}{4\pi(l_g+m)!}}\right)P_{l_g m}(cos\theta)e^{im\varphi} \tag{4.25}$$

Therefore, Laplace's equation describing the CMB can be expressed in terms of the inverse gravity concept l_g such that:

$$\frac{\Phi_m(\varphi)}{sin\theta}\frac{d}{d\theta}\left(sin\theta\frac{d}{d\theta}\right) + \frac{\Theta_{l_g}^m(\theta)}{sin^2\theta}\frac{d^2\Phi_m(\theta)}{d\Phi^2} + l_g(l_g + 1)\Theta_{l_g}^m(\theta)\Phi_m(\varphi) = 0 \tag{4.26}$$

which gives to the angular power spectrum C_{l_g} in terms of the inverse gravity cosmological concept pertaining to the inverse gravity cosmological concept wavenumber l_g such that:

$$C_{l_g} = \frac{2}{2l_g+1}\sum_{m=-1}^{l_g}\left|a_{l_g m}\right|^2 \tag{4.27}$$

42

This concludes a mathematical description of the power spectrum of the cosmic microwave background in terms of the inverse gravity cosmological concept.

Chapter 5

Linearize Gravity and Its Application to the Inverse Gravity Concept

Gravity on the subatomic level or in the quantum realm is described by the mathematics of linearized gravity. Therefore, in chapter 5, I present a mathematical formulation of linearized gravity in terms of the inverse gravity concept. Thus, I present the mathematical structure of linearized gravity based on established concepts and mathematics accepted by the physics community, and then I will apply and integrate the inverse gravity concept into them. Additionally, I present and conclude chapter 5 with an idea of an attractive and repulsive graviton expressed in one of my inverse gravity cosmology papers.

Thus, we begin with the metric tensor $g_{\mu\nu}$ which has a value shown below (Hirata, 2012).

$$g_{\mu\nu} = \eta_{\mu\nu} + h_{\mu\nu} \tag{5.0}$$

where $\eta_{\mu\nu}$ is the Minkowski metric and $h_{\mu\nu}$ the perturbation tensor. The Minkowski metric is expressed such that (Hirata, 2012):

$$\eta_{\mu\nu} = \begin{bmatrix} -1 & 0 & 0 & 0 \\ 0 & 1 & 0 & 0 \\ 0 & 0 & 1 & 0 \\ 0 & 0 & 0 & 1 \end{bmatrix} \tag{5.01}$$

Hence, the perturbation tensor $h_{\mu\nu}$ is expressed such that (Hirata, 2012):

$$h_{\mu\nu} = A_{\mu\nu} e^{ik_\mu x_\mu} \tag{5.02}$$

where the product $k_\mu x_\mu$ in the plane wave expression $e^{ik_\mu x_\mu}$ is the linear combination or superposition of:

$$k_\mu x_\mu = k_0 x_0 + k_1 x_1 + k_2 x_2 + k_3 x_3 \tag{5.03}$$

Here wave number k_μ ($k_\mu \in R^4$) and position x_μ ($x_\mu \in R^4$) are elements in four space, and $A_{\mu\nu}$ is the polarization tensor pertaining to the *transverse traceless gauge* which has a value such that (Hirata, 2012):

$$A_{\mu\nu} = \begin{bmatrix} 0 & 0 & 0 & 0 \\ 0 & A_+ & A_x & 0 \\ 0 & A_x & -A_+ & 0 \\ 0 & 0 & 0 & 0 \end{bmatrix} \tag{5.04}$$

where components A_+ constitute *plus polarization* and components A_x constitute *cross polarization*. The four-component vector valued partial derivatives ∂_μ have a value such that (Hirata, 2012):

$$\partial_\mu = (\partial_0, \partial_1, \partial_2, \partial_3) = \left(\frac{\partial}{\partial x_0}, \frac{\partial}{\partial x_1}, \frac{\partial}{\partial x_2}, \frac{\partial}{\partial x_3}\right) \tag{5.05}$$

Thus, the vector valued partial derivative ∂_μ is multiplied to the perturbation tensor $h_{\mu\nu}$ and set equal to zero as conveyed below (Hirata, 2012).

$$\partial_\mu h_{\mu\nu} = \partial_\mu A_{\mu\nu} e^{ik_\mu x_\mu} = 0 \tag{5.06}$$

where Eq. 5.06, constitutes the *transverse traceless gauge* and adheres to the Lorenz condition. This is alternatively expressed in vector form such that (Hirata, 2012):

$$\partial_\mu h_{\mu\nu} = (0, \ 0, \quad 0, \ 0) \tag{5.07}$$

Eq. 5.07 can alternatively be expressed such that (Hirata, 2012):

$$\partial_\mu h_{\mu\nu} = A_{\mu\nu} e^{ik_\mu x_\mu} (0, (\partial_1 A_+ + \partial_2 A_x), (\partial_1 A_x - \partial_2 A_+), 0) = 0 \tag{5.08}$$

Carrying out the differentiation of vector valued partial derivative ∂_μ gives the expression such that:

$$\partial_\mu h_{\mu\nu} = (0, (ik_1 A_+ + ik_2 A_x), (ik_1 A_x - ik_2 A_+), 0) A_{\mu\nu} e^{ik_\mu x_\mu} = 0 \tag{5.09}$$

The terms must satisfy the Lorenz gauge condition ($\partial_\mu h_{\mu\nu} = 0$). Thus, I present the solutions of:

$$ik_1 A_+ = -ik_2 A_x \qquad (5.10)$$

$$ik_1 A_x = ik_2 A_+ \qquad (5.11)$$

which corresponds to a system of equations such that:

$$(5.12)$$

$$ik_1 A_+ + ik_2 A_x = 0$$

$$ik_1 A_x - ik_2 A_+ = 0$$

Thus, the value of wave component k_1 is a solution to the system of equations corresponding to the top row of the system of equations which is expressed such that:

$$k_1 = -k_2 \frac{A_x}{A_+} \qquad (5.13)$$

Hence, substituting the value of wave component k_1 above into the second (or bottom) row of the system of equations gives:

$$\left(-k_2 \frac{A_x}{A_+} - k_2 A_+ \right) = 0 \qquad (5.14)$$

Eliminating wave number component $-k_2$ from Eq. 5.14 gives:

$$\frac{A_x}{A_+} + A_+ = 0 \qquad (5.15)$$

Thus, the polarization tensor components A_x and A_+ relate to each other via the equation of:

$$A_x = -2A_+ \qquad (5.16)$$

where wave number component k_2 can assume any number value including zero. Thus, table 1 below presents the solution set and its conditions.

Table 1

Solution set:

$$ik_1 A_+ = -ik_2 A_x$$

$$ik_1 A_x = ik_2 A_+$$

For conditions:

$$k_1 = -k_2 \frac{A_x}{A_+}$$

$$\frac{A_x}{A_+} + A_+ = 0$$

Where,

$$A_x = -2A_+$$

k_2 is any arbitrary number

Hence, the solution set of $ik_1 A_+ = -ik_2 A_x$ and $ik_1 A_x = ik_2 A_+$ are applied to $\partial_\mu h_{\mu\nu}$:

$$\partial_\mu h_{\mu\nu} = \left(0, \left((-ik_2 A_x) + ik_2 A_x\right), \left(ik_1 A_x - (ik_1 A_x)\right), 0\right) = 0 \tag{5.17}$$

which reduces to:

$$\partial_\mu h_{\mu\nu} = (0, (0), (0), 0) = 0 \tag{5.18}$$

which corresponds to the condition and expression (Hirata, 2012):

$$k_\mu A_{\mu\nu} = 0 \tag{5.19}$$

thus satisfying the Lorenz condition. We now acknowledge the relation of angular velocity ω_p, energy value E, and reduced Planck's constant $\hbar$ shown below (Hirata, 2012).

$$\omega_p = \frac{E}{\hbar} \tag{5.20}$$

This corresponds to the dispersion relations of:

$$E^2 = k_\mu^2 \tag{5.21}$$

Setting reduced Planck's constant equal to unity ($\hbar = 1$) implies that:

$$\omega_p^2 = \frac{E^2}{\hbar^2} = k_\mu^2 \tag{5.22}$$

which is equivalent to (Hirata, 2012):

$$\omega_p^2 = k_\mu^2 \tag{5.23}$$

In terms of a summation, this is expressed such that:

$$\omega_p^2 = \sum_0^3 k_\mu^2 \tag{5.24}$$

where Eq. 5.24 can be expressed such that:

$$\omega_p^2 = k_0^2 + k_1^2 + k_2^2 + k_3^2 \tag{5.25}$$

Thus, the product of the partial derivative ∂_μ and the Minkowski metric $\eta_{\mu\nu}$ call for raising the indices on the partial derivative of the contravariant 4-vector ∂^μ such that:

$$\partial^\mu = \eta_{\mu\nu}\partial_\mu \tag{5.26}$$

Thus, we have:

$$\partial^\mu = \left(-\frac{\partial}{\partial x_0}, \frac{\partial}{\partial x_1}, \frac{\partial}{\partial x_2}, \frac{\partial}{\partial x_3}\right) \tag{5.27}$$

At this juncture, we apply the contravariant derivative ∂^μ to the product $\partial_\mu h_{\mu\nu}$ which gives:

$$\partial^\mu \partial_\mu h_{\mu\nu} = \partial^\mu \partial_\mu e^{ik_\mu x_\mu} = \partial^\mu \left[A_{\mu\nu} e^{ik_\mu x_\mu} (0, (\partial_1 A_+ + \partial_2 A_x), (\partial_1 A_x - \partial_2 A_+), 0) \right] \tag{5.28}$$

Hence, carrying out the differentiation of the covariant partial derivative ∂_μ gives a value of:

$$\partial_\mu h_{\mu\nu} = \partial^\mu [(0, (ik_1 A_+ + ik_2 A_x), (ik_1 A_x - ik_2 A_+), 0)] = 0 \tag{5.29}$$

Therefore, applying the solutions set of $ik_1 A_+ = -ik_2 A_x$ and $ik_1 A_x = ik_2 A_+$ gives:

$$\partial^\mu \partial_\mu h_{\mu\nu} = \partial^\mu \partial_\mu e^{ik_\mu x_\mu} = \partial^\mu \left[A_{\mu\nu} e^{ik_\mu x_\mu} (0,0,0,0) \right] \tag{5.30}$$

Thus, we apply the contravariant partial derivative ∂^μ such that:

$$\partial^\mu \partial_\mu h_{\mu\nu} = \partial^\mu \left(\partial_\mu A_{\mu\nu} e^{ik_\mu x_\mu} \right) = \left[A_{\mu\nu} e^{ik_\mu x_\mu} ((0, (\partial_1 A_+ + \partial_2 A_x), (\partial_1 A_x - \partial_2 A_+), 0))(0,0,0,0) \right] \tag{5.31}$$

And thus, carrying out the differentiation corresponding to contravariant partial derivative ∂^μ gives:

$$\partial^\mu \partial_\mu h_{\mu\nu} = \partial^\mu \left(\partial_\mu A_{\mu\nu} e^{ik_\mu x_\mu}\right) = A_{\mu\nu} e^{ik_\mu x_\mu}\left((0, (ik_1 A_+ + ik_2 A_x), (ik_1 A_x - ik_2 A_+), 0)\right)(0,0,0,0)$$

(5.32)

Once more applying the solution set:

$$ik_1 A_+ = -ik_2$$

(5.33)

$$ik_1 A_x = ik_2 A_+$$

gives the result:

$$\partial^\mu \partial_\mu h_{\mu\nu} = A_{\mu\nu} e^{ik_\mu x_\mu}\left((0,0,0,0)\right)(0,0,0,0) = 0$$

(5.34)

which corresponds to the condition and expression (Hirata, 2012):

$$k^\mu k_\mu A_{\mu\nu} = 0$$

(5.35)

thus, embodying the Lorenz condition of (Hirata, 2012):

$$\partial^\mu \partial_\mu h_{\mu\nu} = 0$$

(5.36)

Eq. 5.36 constitutes the Laplacian of the form (Hirata, 2012):

$$\nabla_0^2 \Phi + \nabla_1^2 \Phi + \nabla_2^2 \Phi + \nabla_3^2 \Phi = 0$$

(5.37)

The Laplacian $(\partial_\mu \partial^\mu h_{\mu\nu} = 0)$ constitutes the d'Alembertian operator $\square$ which is of the form (Hirata, 2012):

$$\Box \, \Phi = \partial^\mu \partial_\mu \Phi = \frac{\partial^2 \Phi}{\partial x_1^2} + \frac{\partial^2 \Phi}{\partial x_2^2} + \frac{\partial^2 \Phi}{\partial x_3^2} - \frac{\partial^2 \Phi}{\partial x_0^2} = 0 \qquad (5.38)$$

where c is the velocity of light and t is time. Coordinate x_0 in the d'Alembert Ian operator above has a value such that:

$$x_0 = ct \qquad (5.39)$$

Thus, applying the d'Alembert Ian operator to the perturbation tensor $h_{\mu\nu}$ gives as shown below (Hirata, 2012).

$$\Box \, h_{\mu\nu} = \partial^\mu \partial_\mu h_{\mu\nu} \qquad (5.40)$$

Consider the Stress-energy tensor $T_{\mu\nu}$ which has a value as shown below (Hirata, 2012).

$$T_{\mu\nu} = \rho \partial_\mu \Phi \partial_\nu \Phi \qquad (5.41)$$

The stress-energy tensor $T_{\mu\nu}$ is equal to the product of the d'Alembertian operator $\Box$ and the metric tensor $g_{\mu\nu}$ as shown below (Hirata, 2012).

$$T_{\mu\nu} = \Box \, g_{\mu\nu} \qquad (5.42)$$

where

$$\Box \, g_{\mu\nu} = \partial^\mu \partial_\mu g_{\mu\nu} \qquad (5.43)$$

which shows that:

$$\Box \, g_{\mu\nu} = \partial^\mu \partial_\mu \big(\eta_{\mu\nu} + h_{\mu\nu} \big) \qquad (5.44)$$

Thus, distributing the product of differentials $\partial^\mu \partial_\mu$ gives an expression such that:

$$\Box\, g_{\mu\nu} = \partial^\mu \partial_\mu g_{\mu\nu} = \partial^\mu \partial_\mu \eta_{\mu\nu} + \partial^\mu \partial_\mu h_{\mu\nu} \tag{5.45}$$

The Minkowski metric vanishes under differentiation, giving:

$$\Box\, g_{\mu\nu} = \left[0 + \partial^\mu \partial_\mu h_{\mu\nu}\right] = \partial^\mu \partial_\mu h_{\mu\nu} \tag{5.46}$$

which implies that:

$$\Box\, g_{\mu\nu} = \partial^\mu \partial_\mu g_{\mu\nu} = \partial^\mu \partial_\mu h_{\mu\nu} \tag{5.47}$$

Thus, the stress-energy tensor relates to the perturbation tensor $h_{\mu\nu}$ such that (Hirata, 2012):

$$T_{\mu\nu} = \partial^\mu \partial_\mu h_{\mu\nu} \tag{5.48}$$

Where ρ is the correlating density and Φ the correlating field function, the stress tensor $T_{\mu\nu}$ is expressed in its 4 by 4 matrix form such that (Hirata, 2012):

$$T_{\mu\nu} = \rho \begin{bmatrix} \partial_0\Phi\partial_0\Phi & \partial_0\Phi\partial_1\Phi & \partial_0\Phi\partial_2\Phi & \partial_0\Phi\partial_3\Phi \\ \partial_1\Phi\partial_0\Phi & \partial_1\Phi\partial_1\Phi & \partial_1\Phi\partial_2\Phi & \partial_1\Phi\partial_3\Phi \\ \partial_2\Phi\partial_0\Phi & \partial_2\Phi\partial_1\Phi & \partial_2\Phi\partial_2\Phi & \partial_2\Phi\partial_3\Phi \\ \partial_3\Phi\partial_0\Phi & \partial_3\Phi\partial_1\Phi & \partial_3\Phi\partial_2\Phi & \partial_3\Phi\partial_3\Phi \end{bmatrix} \tag{5.49}$$

The product of differentials $\partial_\mu \partial^\mu$ can be expressed as a summation as shown below.

$$\partial^\mu \partial_\mu = \sum_0^3 \partial^\mu \partial_\mu \tag{5.50}$$

Thus, applying this to the perturbation tensor $h_{\mu\nu}$ gives the equation of:

$$\partial^\mu \partial_\mu h_{\mu\nu} = \sum_0^3 \partial^\mu \partial_\mu \, h_{\mu\nu} \tag{5.51}$$

Therefore, expressing this as an expansion gives:

$$\partial^\mu \partial_\mu h_{\mu v} = \partial^0 \partial_0 h_{\mu v} + \partial^1 \partial_1 h_{\mu v} + \partial^2 \partial_2 h_{\mu v} + \partial^3 \partial_3 h_{\mu v} \tag{5.52}$$

The product of $\partial^\mu \partial_\mu h_{\mu v}$ can be expressed as a sum of 4 by 4 matrices such that:

$$\partial^\mu \partial_\mu h_{\mu v} = \begin{bmatrix} 0 & 0 & 0 & 0 \\ 0 & 0 & 0 & 0 \\ 0 & 0 & 0 & 0 \\ 0 & 0 & 0 & 0 \end{bmatrix} + \begin{bmatrix} 0 & 0 & 0 & 0 \\ 0 & -k_1^2 A_+ e^{ik_\mu x_\mu} & -k_1^2 A_x e^{ik_\mu x_\mu} & 0 \\ 0 & -k_1^2 A_x e^{ik_\mu x_\mu} & k_1^2 A_+ e^{ik_\mu x_\mu} & 0 \\ 0 & 0 & 0 & 0 \end{bmatrix} +$$

$$\begin{bmatrix} 0 & 0 & 0 & 0 \\ 0 & -k_2^2 A_+ e^{ik_\mu x_\mu} & -k_2^2 A_x e^{ik_\mu x_\mu} & 0 \\ 0 & -k_2^2 A_x e^{ik_\mu x_\mu} & k_2^2 A_+ e^{ik_\mu x_\mu} & 0 \\ 0 & 0 & 0 & 0 \end{bmatrix} + \begin{bmatrix} 0 & 0 & 0 & 0 \\ 0 & -k_3^2 A_+ e^{ik_\mu x_\mu} & -k_3^2 A_x e^{ik_\mu x_\mu} & 0 \\ 0 & -k_3^2 A_x e^{ik_\mu x_\mu} & k_3^2 A_+ e^{ik_\mu x_\mu} & 0 \\ 0 & 0 & 0 & 0 \end{bmatrix} \tag{5.53}$$

Combining the sum of matrices gives:

$$\partial^\mu \partial_\mu h_{\mu v} = \begin{bmatrix} 0 & 0 & 0 & 0 \\ 0 & -\sum_1^3 k_\mu^2 A_+ e^{ik_\mu x_\mu} & -\sum_1^3 k_\mu^2 A_x e^{ik_\mu x_\mu} & 0 \\ 0 & -\sum_1^3 k_\mu^2 A_x e^{ik_\mu x_\mu} & \sum_1^3 k_\mu^2 A_+ e^{ik_\mu x_\mu} & 0 \\ 0 & 0 & 0 & 0 \end{bmatrix} \tag{5.54}$$

where the components $\pm \sum_1^3 k_\mu^2 A_a e^{ik_\mu x_\mu}$ give an expansion for each component of the polarization tensor $A_a (A_x$ and/or $A_+)$, expressed such that:

$$\sum_1^3 k_\mu^2 A_a e^{ik_\mu x_\mu} = k_0^2 A_a e^{ik_0 x_0} + k_1^2 A_a e^{ik_1 x_1} + k_2^2 A_a e^{ik_2 x_2} + k_3^2 A_a e^{ik_3 x_3} \tag{5.55}$$

Recall the equivalence of Laplacian $\partial^\mu \partial_\mu h_{\mu v}$ and the stress energy tensor $T_{\mu v}$ shown below (Wald, 1984).

$$\partial^\mu \partial_\mu h_{\mu v} = T_{\mu v} \tag{5.56}$$

Hence, in matrix form, this equivalence can be expressed as:

$$
\begin{bmatrix}
0 & 0 & 0 & 0 \\
0 & -\sum_1^3 k_\mu^2 A_+ e^{ik_\mu x_\mu} & -\sum_1^3 k_\mu^2 A_x e^{ik_\mu x_\mu} & 0 \\
0 & -\sum_1^3 k_\mu^2 A_x e^{ik_\mu x_\mu} & \sum_1^3 k_\mu^2 A_+ e^{ik_\mu x_\mu} & 0 \\
0 & 0 & 0 & 0
\end{bmatrix}
= \rho
\begin{bmatrix}
\partial_0 \Phi \partial_0 \Phi & \partial_0 \Phi \partial_1 \Phi & \partial_0 \Phi \partial_2 \Phi & \partial_0 \Phi \partial_3 \Phi \\
\partial_1 \Phi \partial_0 \Phi & \partial_1 \Phi \partial_1 \Phi & \partial_1 \Phi \partial_2 \Phi & \partial_1 \Phi \partial_3 \Phi \\
\partial_2 \Phi \partial_0 \Phi & \partial_2 \Phi \partial_1 \Phi & \partial_2 \Phi \partial_2 \Phi & \partial_2 \Phi \partial_3 \Phi \\
\partial_3 \Phi \partial_0 \Phi & \partial_3 \Phi \partial_1 \Phi & \partial_3 \Phi \partial_2 \Phi & \partial_3 \Phi \partial_3 \Phi
\end{bmatrix}
$$

(5.57)

Thus, the component of each matrix has a value such that:

$$ -\sum_1^3 k_\mu^2 A_+ e^{ik_\mu x_\mu} = \rho \partial_1 \Phi \partial_1 \Phi $$

(5.58)

$$ -\sum_1^3 k_\mu^2 A_x e^{ik_\mu x_\mu} = \rho \partial_2 \Phi \partial_1 \Phi $$

(5.59)

$$ -\sum_1^3 k_\mu^2 A_x e^{ik_\mu x_\mu} = \rho \partial_1 \Phi \partial_2 \Phi $$

(5.60)

$$ \sum_1^3 k_\mu^2 A_+ e^{ik_\mu x_\mu} = \rho \partial_2 \Phi \partial_2 \Phi $$

(5.61)

Hence, the stress energy tensor $T_{\mu v}$ can be expressed such that:

$$ T_{\mu v} = \partial^\mu \partial_\mu h_{\mu v} = \rho \partial_\mu \Phi \partial_v \Phi $$

(5.62)

or equivalently,

$$ T_{\mu v} = \rho \partial_\mu \Phi \partial_v \Phi = \pm k_\mu k^\mu A_{\mu v} e^{ik_\mu x_\mu} $$

(5.63)

Here I present a new solution for field function Φ assigning it a value as shown below.

$$ \Phi = \frac{2\hbar}{m_0} b_w^{1/2} e^{i\frac{k_\mu x_\mu}{2}} $$

(5.64)

Where b_w is a constant of proportionality, wave number k_μ can be expressed such that (Young& Freedman, 2004):

$$k_\mu = \frac{p_\mu}{\hbar}$$
(5.65)

Momentum value p_μ can be expressed in terms of an arbitrary mass m_0 such that (Young & Freedman, 2004):

$$p_\mu = m_0 u_\mu$$
(5.66)

Where u_μ is a four velocity ($u_\mu \in R^4 \cup C$ *where C is the set of imaginary numbers*), wave number k_μ is expressed such that:

$$k_\mu = \left(\frac{m_0 u_\mu}{\hbar}\right)$$
(5.67)

Thus, the four velocity u_μ can be expressed such that:

$$u_\mu = \frac{\hbar}{m_0} k_\mu = \frac{\hbar}{m_0}\left(\frac{m_0 u_\mu}{\hbar}\right)$$
(5.68)

With the establishment of four velocity u_μ, we briefly transition to the field function Φ prior to returning to four velocity u_μ. Thus, substituting the value of field function Φ into the stress-energy tensor $\rho \partial_\mu \Phi \partial_\nu \Phi$ gives:

$$\rho \partial_\mu \Phi \partial_\nu \Phi = \rho \partial_\mu \left(\frac{2\hbar}{m_0} b_w^{1/2} e^{i\frac{k_\mu x_\mu}{2}}\right) \partial_\nu \left(\frac{2\hbar}{m_0} b_w^{1/2} e^{i\frac{k_\mu x_\mu}{2}}\right)$$
(5.69)

Thus, carrying out the partial differentiation of partial derivatives ∂_a, the value of the stress energy tensor $\rho \partial_\mu \Phi \partial_\nu \Phi$ in terms of four velocity u_μ is such that:

$$\rho \partial_\mu \Phi \partial_\nu \Phi = \rho \left(iu_\mu b_w^{1/2} e^{i\frac{k_\mu x_\mu}{2}}\right)\left(iu_\nu b_w^{1/2} e^{i\frac{k_\mu x_\mu}{2}}\right)$$
(5.70)

Whereas previously expressed,

$$\frac{2\hbar}{m_0} k_\mu = u_\mu \tag{5.71}$$

Hence, the stress-energy tensor can be expressed such that:

$$\rho \partial_\mu \Phi \partial_\nu \Phi = -\rho u_\mu u_\nu b_w e^{ik_\mu x_\mu} \tag{5.72}$$

where the standard form of the stress energy tensor is (Wald, 1984):

$$\rho \partial_\mu \Phi \partial_\nu \Phi = -\rho u_\mu u_\nu \tag{5.73}$$

The components of the stress-energy tensor have equivalent values such that:

$$-\sum_1^3 k_\mu^2 A_+ e^{ik_\mu x_\mu} = \rho \partial_1 \Phi \partial_1 \Phi \equiv -\rho u_1 u_1 b_w e^{ik_\mu x_\mu} \tag{5.74}$$

$$-\sum_1^3 k_\mu^2 A_x e^{ik_\mu x_\mu} = \rho \partial_2 \Phi \partial_1 \Phi \equiv -\rho u_2 u_1 b_w e^{ik_\mu x_\mu} \tag{5.75}$$

$$-\sum_1^3 k_\mu^2 A_x e^{ik_\mu x_\mu} = \rho \partial_1 \Phi \partial_2 \Phi \equiv -\rho u_1 u_2 b_w e^{ik_\mu x_\mu} \tag{5.76}$$

$$\sum_1^3 k_\mu^2 A_+ e^{ik_\mu x_\mu} = \rho \partial_2 \Phi \partial_2 \Phi \equiv -\rho u_2 u_2 b_w e^{ik_\mu x_\mu} \tag{5.77}$$

In four-by-four matrix form, this is expressed as:

$$\begin{bmatrix} 0 & 0 & 0 & 0 \\ 0 & -\sum_1^3 k_\mu^2 A_+ e^{ik_\mu x_\mu} & -\sum_1^3 k_\mu^2 A_x e^{ik_\mu x_\mu} & 0 \\ 0 & -\sum_1^3 k_\mu^2 A_x e^{ik_\mu x_\mu} & \sum_1^3 k_\mu^2 A_+ e^{ik_\mu x_\mu} & 0 \\ 0 & 0 & 0 & 0 \end{bmatrix} = \rho \begin{bmatrix} 0 & 0 & 0 & 0 \\ 0 & -\rho u_1 u_1 b_w e^{ik_\mu x_\mu} & -\rho u_1 u_2 b_w e^{ik_\mu x_\mu} & 0 \\ 0 & -\rho u_2 u_1 b_w e^{ik_\mu x_\mu} & -\rho u_2 u_2 b_w e^{ik_\mu x_\mu} & 0 \\ 0 & 0 & 0 & 0 \end{bmatrix} \tag{5.78}$$

Therefore, at this juncture, we determine the values of each component of four velocity u_μ which are velocity components u_1 and u_2 presented in the following.

$$-\sum_1^3 k_\mu^2 A_+ e^{ik_\mu x_\mu} = -\rho u_1^2 b_w e^{ik_\mu x_\mu} \tag{5.79}$$

with a velocity component u_1 value of:

$$u_1 = \left[(\rho b_w)^{-1} \sum_1^3 k_\mu^2 A_+\right]^{1\backslash 2} \tag{5.80}$$

And for the equation containing velocity component u_2 such that:

$$\sum_1^3 k_\mu^2 A_+ e^{ik_\mu x_\mu} = -\rho u_2^2 e^{ik_\mu x_\mu} \tag{5.81}$$

velocity component u_2 has a value of:

$$u_2 = i\left[(\rho b_w)^{-1} \sum_1^3 k_\mu^2 A_+\right]^{1\backslash 2} \tag{5.82}$$

Hence, velocity components u_0 and u_3 equal zero as shown below.

$$u_0 = u_3 = 0 \tag{5.83}$$

Thus, we substitute the values of components u_1 and u_2 into the equation of:

$$-\sum_1^3 k_\mu^2 A_x e^{ik_\mu x_\mu} = -\rho u_2 u_1 b_w e^{ik_\mu x_\mu} \tag{5.84}$$

which gives the result:

$$-\sum_1^3 k_\mu^2 A_x e^{ik_\mu x_\mu} = -\rho \left[i\left[(\rho b_w)^{-1} \sum_1^3 k_\mu^2 A_+\right]^{1\backslash 2}\right]\left[\left[(\rho b_w)^{-1} \sum_1^3 k_\mu^2 A_+\right]^{1\backslash 2}\right] e^{ik_\mu x_\mu} \tag{5.85}$$

This reduces to:

$$-\sum_1^3 k_\mu^2 A_x e^{ik_\mu x_\mu} = i\rho(\rho b_w)^{-1} \sum_1^3 k_\mu^2 A_+ e^{ik_\mu x_\mu} \tag{5.86}$$

which implies that:

$$-\sum_1^3 k_\mu^2 A_x = i\sum_1^3 k_\mu^2 b_w A_+ \tag{5.87}$$

where constant b_w has a value such that:

$$b_w = -\frac{A_x}{iA_+} \tag{5.88}$$

Hence, the field function $\varPhi$ can be expressed as shown below.

$$\varPhi = \frac{2\hbar i}{m_0}\left(\frac{A_x}{iA_+}\right)^{1/2} e^{i\frac{k_\mu x_\mu}{2}} \tag{5.89}$$

This completes the solution and mathematical exposition of linearized gravity. At this juncture, I show the incorporation of the inverse gravity concept. Recall the value of gravitational potential energy $U_T(r)$ in terms of the inverse gravity concept shown below.

$$U_T(r) = \left[\frac{1}{r_0^2}\right]\left[\frac{r^3}{3GM_A'M_B'}\right] + \frac{GM_A'M_B'}{r} + W_r \tag{5.90}$$

Hence, we set inverse gravity concept potential energy $U_T(r)$ equal to relativistic energy pc, which is the product of momentum p and velocity of light c as conveyed below.

$$U_T(r) = pc \tag{5.91}$$

Hence, momentum p can be expressed such that:

$$p = \frac{U_g(r)}{c} \tag{5.92}$$

Thus, in terms of the inverse gravity concept, momentum p takes on a value such that:

$$p = \frac{1}{c}\left[\left[\frac{1}{r_0^2}\right]\left[\frac{r^3}{3GM_A'M_B'}\right] + \frac{GM_A'M_B'}{r} + W_r\right] \tag{5.93}$$

or alternatively,

$$p = \left[\left[\frac{1}{cr_0^2} \right] \left[\frac{r^3}{3GM_A'M_B'} \right] + \frac{GM_A'M_B'}{cr} + \frac{W_r}{c} \right] \tag{5.94}$$

Extending momentum p into 4-space, we have 4 space momentum p_μ which is expressed such that:

$$p_\mu = pa_\mu \tag{5.95}$$

Where a_μ is a directional unit vector such that $|a_\mu| = 1$ and has a value such that:

$$a_\mu = (a_0, \ a_1, \ a_2, \ a_3) \tag{5.96}$$

Therefore, 4 space momentum value p_μ has a value such that:

$$p_\mu = pa_\mu \equiv (pa_0, \ pa_1, pa_2 \ , pa_3) \tag{5.97}$$

Thus, in line with the specific mathematical structure of linearized gravity presented in this chapter (and therefore book), momentum p_μ has a value such that:

$$p_\mu = (0, \ pa_1, \ pa_2 \ , 0) \tag{5.98}$$

In terms of the inverse gravity concept, momentum p_μ can be expressed such that:

$$p_\mu = a_\mu \left[\left[\frac{1}{cr_0^2} \right] \left[\frac{r^3}{3GM_A'M_B'} \right] + \frac{GM_A'M_B'}{cr} + \frac{W_r}{c} \right] \tag{5.99}$$

Momentum relates to wave number k_μ via the equation of:

$$k_\mu = \frac{p_\mu}{\hbar} \tag{5.100}$$

where the wave number $k_{g\mu}$ denotes wave number in terms of the inverse gravity concept. Thus wave number $k_{g\mu}$ has a value such that:

$$k_{g\mu} = \frac{a_\mu}{\hbar}\left[\left[\frac{1}{cr_0^2}\right]\left[\frac{r^3}{3GM_A'M_B'}\right] + \frac{GM_A'M_B'}{cr} + \frac{W_r}{c}\right] \qquad (5.101)$$

The solution scalar field Φ which, once more, has a value such that:

$$\Phi = \frac{2\hbar i}{m_0}\left(\frac{A_x}{iA_+}\right)^{1/2} e^{i\frac{k_{g\mu}x_\mu}{2}} \qquad (5.102)$$

can be expressed in terms of the inverse gravity concept such that:

$$\Phi = \frac{2\hbar i}{m_0}\left(\frac{A_x}{iA_+}\right)^{1/2} exp\left(\frac{ix_\mu}{2}\left[\frac{a_\mu}{\hbar}\left[\left[\frac{1}{cr_0^2}\right]\left[\frac{r^3}{3GM_A'M_B'}\right] + \frac{GM_A'M_B'}{cr} + \frac{W_r}{c}\right]\right]\right) \qquad (5.103)$$

This implies that the inverse gravity wave number component k_{g1} must adhere to the solution set condition of:

$$k_{g1} = -k_{g2}\frac{A_x}{A_+} \qquad (5.104)$$

where inverse gravity wave number component k_{g2} can assume any value. This can alternatively be expressed such that:

$$\frac{a_1}{\hbar}\left[\left[\frac{1}{cr_0^2}\right]\left[\frac{r^3}{3GM_A'M_B'}\right] + \frac{GM_A'M_B'}{cr} + \frac{W_r}{c}\right] = -\frac{a_2}{\hbar}\left[\left[\frac{1}{cr_0^2}\right]\left[\frac{r^3}{3GM_A'M_B'}\right] + \frac{GM_A'M_B'}{cr} + \frac{W_r}{c}\right]\frac{A_x}{A_+} \qquad (5.105)$$

In my paper titled "The Relationship between the Cosmological Inverse Gravity Assertion and the Cosmological Constant including an Alternative Possibility of the Graviton," the inverse gravity concept was applied to the notion of a spin 2 graviton whose description lies within the context of linearized gravity. The possibility of attractive and/or repulsive force within a graviton was introduced. The graviton is a massless particle and speculative in nature at the time of writing. Gravitons are required to have long ranges due to the extensive range of gravity and are suspected to propagate at the velocity of light. Moreover, the graviton, as stated, must be a spin-2 boson because the source is the stress-energy tensor as we showed ($T_{\mu\nu} = \rho\partial_\mu\Phi\partial_\nu\Phi$). The perturbation tensor $h_{\mu\nu}$ constitutes a wave amplitude; hence, we once more apply the d'Alembert Ian operator $\square$ to the perturbation tensor $h_{\mu\nu}$ as shown below (Walker, 2018).

$$\square\, h_{\mu\nu} = \partial_\mu\partial^\mu h_{\mu\nu} \tag{5.106}$$

which can equivalently be expressed such that:

$$\partial_\mu\partial^\mu h_{\mu\nu} = -\sum_1^3 k_\mu^2\, A_{\mu\nu} e^{ik_\mu x_\mu} \tag{5.107}$$

where once more, the perturbation tensor has a value such that:

$$h_{\mu\nu} = -A_{\mu\nu} e^{ik_\mu x_\mu} \tag{5.108}$$

Thus, applying the Laplacian gives:

$$\partial_\mu\partial^\mu h_{\mu\nu} = -\sum_1^3 k_\mu^2 h_{\mu\nu} \tag{5.109}$$

Thus, k_μ^2 equals the Pythagorean expansion of:

$$k_\mu^2 = \sum_1^3 k_\mu^2 \tag{5.110}$$

This gives:

$$\partial_\mu\partial^\mu h_{\mu\nu} = -k_\mu^2 h_{\mu\nu} \tag{5.111}$$

Applying the coefficient of $\hbar^2 c^2$ gives an Eigen value of energy E such that:

$$(\hbar^2 c^2)\partial_\mu\partial^\mu h_{\mu\nu} = -(\hbar^2 c^2)k_\mu^2 h_{\mu\nu} \equiv -E h_{\mu\nu} \tag{5.112}$$

This implies that:

$$-(\hbar^2 c^2) k_\mu^2 h_{\mu v} \equiv -E h_{\mu v} \tag{5.113}$$

which reduces to:

$$(\hbar^2 c^2) k_\mu^2 \, h_{\mu v} \equiv E h_{\mu v} \tag{5.114}$$

Eliminating perturbation tensor $h_{\mu v}$ from both sides of the equation gives the energy E a value such that:

$$E = (\hbar^2 c^2) k_{g\mu}^2 \tag{5.115}$$

Substituting the value of inverse gravity wave number $k_{g\mu}$ into the equation above gives energy E in terms of the inverse gravity assertion such that:

$$E = \Sigma_1^3 \left[\left[\frac{a_\mu}{r_0^2} \right] \left[\frac{r^3}{3GM_A'M_B'} \right] + \frac{GM_A'M_B'}{r} + W_r \right]^2 \tag{5.116}$$

In theory, the energy of a graviton would be repulsive for the condition of:

$$\left[\frac{a_\mu}{r_0^2} \right] \left[\frac{r^3}{3GM_A'M_B'} \right] > \frac{GM_A'M_B'}{r} \tag{5.117}$$

and attractive for the condition of:

$$\left[\frac{a_\mu}{r_0^2} \right] \left[\frac{r^3}{3GM_A'M_B'} \right] < \frac{GM_A'M_B'}{r} \tag{5.118}$$

This shows the theoretical potential of a graviton having the characteristic of being an attractive and/or repulsive particle contingent on distance r between cosmological masses and its correlation to linearized gravity. Thus, we have a holistic description of linearized gravity and its correlation to the inverse gravity cosmological concept.

Chapter 6

The Inverse Gravity Concept Applied to Weakly Interacting Massive Particles as a Cause of Cosmological Expansion

Chapter 6 conveys the incorporation of the inverse gravity concept to weakly interacting massive particles or WIMPs. Weakly interacting massive particles are theorized to be the cause of cosmological expansion. WIMPs are speculated by some physicists to be elusive, heavy, slow moving, and electromagnetically neutral subatomic particles. Furthermore, some physicists suppose that WIMPs are nonbaryonic. Furthermore, WIMPs are hypothesized to be the composition of dark matter which is believed to account for 22 percent of the universe's content (Brittanica.com, 2023).

I assert that weakly interacting massive particles can possibly be described in the realm of electroweak interactions. Thus, in *description 3* of this chapter, the inverse gravity concept is described in terms of electroweak interactions. Weakly interacting particles, in general, are particles that have weak interactions which are governed by the decay of unstable subatomic particles. Weak interactions are conducted or mediated by the W^-, W^+, and Z^0 bosons which are spin 1 particles and have mass values of 80.4 GeV/c^2 and 91.2.4 GeV/c^2 (Young & Freedman, 2004).

Thus, we begin the incorporation of the inverse gravity cosmology concept to weak interactions. We start with the spatial position x'_u and x_u (x_u, $x'_u \in R^3$) in three-dimensional space. Spatial positions x'_u and x_u are the transition points of particle/antiparticle pairs that signify where the particles are created and/or destroyed (or annihilated) producing a mediating virtual particle (i.e., W^-, W^+ and Z^0 bosons). The range of weak interactions are approximately 0.001 femtometers (fm), which we will revisit shortly. The distance between the two creation/annihilation positions x'_u and x_u will be denoted l_{med}. Hence, distance l_{med} is the magnitude of the distance between creation/annihilation positions x'_u and x_u shown below (Walker, 2017).

$$l_{med} = |x'_{\mathrm{u}} - x_{\mathrm{u}}| = \left[\sum_{\mathrm{u}=1}^{\mathrm{u}=3}(x'_{\mathrm{u}} - x_{\mathrm{u}})^2\right]^{1/2} \qquad (6.0)$$

Therefore, distance l_{med} between creation/annihilation positions x'_{u} and x_{u} is less than or equal to 0.001 fm as shown by the inequality below (Walker, 2017).

$$0.001 \text{ fm} \geq l_{med} \qquad (6.01)$$

Recall that the total gravitational energy value $U_T(r)$ in terms of the inverse gravity cosmological concept is given by (Walker, 2017):

$$U_T(r) = \left[\frac{1}{r_0^2}\right]\left[\frac{r^3}{3GM'_A M'_B}\right] + \frac{GM'_A M'_B}{r} + W_r \qquad (6.02)$$

In this chapter, we are interested in the expansion term which is highlighted in yellow in the equation above. Thus, dark energy E_{dark} is equal to this term as shown below (Walker, 2017).

$$E_{dark} = \left[\frac{1}{r_0^2}\right]\left[\frac{r^3}{3GM'_A M'_B}\right] \qquad (6.03)$$

Energy E_{dark} is a function of cosmological level distance r; therefore, energy E_{dark} is a close approximation to the j (j assumes values of positive integers) sums of interaction energies E_{medj} of the mediating particles (or the W^-, W^+, or Z^0 bosons). Moreover, j is the number of beta decay interactions within a given region of space, which cumulatively constitute the force and thus pressure of cosmological expansion. Thus, in my original paper titled "Cosmology: The Theoretical Possibility of Inverse Gravity as a Cause of Cosmological Inflation in an Isotropic and Homogeneous Universe and Its Relationship to Weakly Interacting Massive Particles," energy E_{dark} relates to the j sums of interaction energies E_{medj} such that (Walker, 2017):

$$E_{dark} \approx \sum_j E_{medj} \qquad (6.04)$$

Thus, a single or individual value of interaction energy E_{med} (or the energy of the mediating W^-, W^+, and Z^0 bosons) is the inverse gravity term in terms of distance l_{med}. Therefore, at the subatomic level, the distance of weak interactions is that of l_{med}; thus, distance r is substituted with distance l_{med} $(r \rightarrow l_{med})$ in the inverse term, giving the energy of the weak interaction mediating particle in terms of distance l_{med} such that:

$$E_{med} = \left[\frac{1}{r_0^2}\right]\left[\frac{(l_{med})^3}{3GM_A'M_B'}\right] \qquad (6.05)$$

Or alternatively,

$$E_{med} = \left[\frac{1}{r_0^2}\right]\left[\frac{|x_u'-x_u|^3}{3GM_A'M_B'}\right] \equiv pc \qquad (6.06)$$

However, this expression is incorrect and requires a modification for the condition of $E_{dark} \approx \sum_j E_{medj}$. Where r is the cosmological distance of the effect of dark energy E_{dark} and 0.001 fm the range of weak interactions, cosmological distance r can be expressed as the product of:

$$r = (0.001 \text{ fm})N_c \qquad (6.07)$$

Hence, real number N_c takes on a value such that:

$$N_c = \frac{r}{0.001 \text{ fm}} \qquad (6.08)$$

Thus, I correct the expression of energy E_{med} (which is not expressed in the original papers) to the expression of:

$$E_{med} = \left[\frac{\alpha_c}{N_c r_0^2}\right]\left[\frac{(l_{med})^3}{3GM_A'M_B'}\right] \qquad (6.09)$$

which incorporates real number N_c and α_c, which is a proportionality constant for the equivalence of:

$$E_{dark} = \left[\frac{1}{r_0^2}\right]\left[\frac{r^3}{3GM_A'M_B'}\right] = \sum_j^{N_c} E_{medj} \qquad (6.10)$$

Alternatively,

$$\left[\frac{1}{r_0^2}\right]\left[\frac{r^3}{3GM_A'M_B'}\right] = \sum_j^{N_c}\left[\left[\frac{\alpha_c}{N_c r_0^2}\right]\left[\frac{(l_{med})^3}{3GM_A'M_B'}\right]\right]_j \qquad (6.11)$$

Thus, dark energy E_{dark} is a compilation (or summation) of N_c number of weak interactions of energy E_{med} in a region of space. The mediating energy E_{med} of the W^-, W^+, or Z^0 bosons of the decay of unstable subatomic particles must obey the Heisenberg uncertainty principle for energy (ΔE) and time (Δt). The Heisenberg uncertainty principle is expressed such that (Young & Freedman, 2004):

$$\Delta E \Delta t \geq \hbar \qquad (6.12)$$

Thus, the uncertainty in energy ΔE can be set equal to inverse gravity mediating energy E_{med} as expressed below (Walker, 2017).

$$\Delta E = E_{med} \qquad (6.13)$$

Thus, equivalently,

$$\Delta E = \left[\frac{\alpha_c}{N_c r_0^2}\right]\left[\frac{(l_{med})^3}{3GM_A'M_B'}\right] \qquad (6.14)$$

As previously stated, the maximum value of length l_{med} is 0.001 fm for weak interactions. Therefore, length l_{med} is set equal to 0.001 fm ($l_{med} = 0.001$ fm), conclusively energy ΔE can be expressed such that (Walker, 2017):

$$\Delta E = \left[\frac{\alpha_c}{N_c r_0^2}\right]\left[\frac{(0.001\ \text{fm})^3}{3GM_A'M_B'}\right] \qquad (6.15)$$

The product of energy (ΔE) and time (Δt) ($\Delta E \Delta t$) can be expressed such that (Walker, 2017):

$$\Delta E \Delta t = \left[\frac{\alpha_c}{N_c r_0^2}\right]\left[\frac{(0.001 \text{ fm})^3}{3GM_A' M_B'}\right]\Delta t \qquad (6.16)$$

The uncertainty principle can be expressed in terms of the inverse gravity concept such that (Walker, 2017):

$$\left[\frac{\alpha_c}{N_c r_0^2}\right]\left[\frac{(0.001 \text{ fm})^3}{3GM_A' M_B'}\right]\Delta t \geq \hbar \qquad (6.17)$$

Here, time span Δt must be sufficiently large to satisfy the inequality above. Thus, the larger the value of the continuous sums of cosmological masses $M_A' M_B'$, the larger time span Δt during which the mediating particle of the decay process can exist according to the Heisenberg uncertainty principle. Moreover, the larger the continuous sums of cosmological masses $M_A' M_B'$, the smaller the energy E_{med} of the W^-, W^+, or Z^0 bosons of the weak interaction in terms of the inverse gravity concept. Hence, the force and energy values of an inverse gravity weak interaction are very minute and very difficult to detect; however, the time span during which they can exist may be longer than that of typical weak interactions according to the uncertainty principle (Young & Freedman, 2004).

The explanation of the expansion of the geometry of space-time in the quantum realm due to dark energy in terms of the inverse gravity concept that included a linearized version of the isotropic Friedman-Walker-Robertson metric for flat space presented in the original paper has been omitted from this book. This is due to the explanation of linearized gravity in terms of the inverse gravity concept presented in chapter 5—not to mention that, by my assessment, it was faulty.

Since the nature of a theoretical weakly interacting particle as it pertains to dark energy and thus cosmological expansion has not been determined and is speculative at the time of this writing, I give three mathematical descriptions of weak interactions that incorporate the inverse gravity concept.

Description 1: Transition amplitudes of weak interactions between particle/antiparticle pairs using an inverse gravity concept mediating or virtual particle.

Thence, diving into quantum field theory, the Dyson expansion of the S operator is expressed such that (Klauber, 2013):

$$S = \sum_{n=0}^{\infty} \frac{(-i)^n}{n!} \int_{-\infty}^{\infty} \cdots \cdots \int_{-\infty}^{\infty} T\left\{H_{\zeta}^{\zeta}(t_1) H_{\zeta}^{\zeta}(t_2) \ldots \ldots H_{\zeta}^{\zeta}(t_n)\right\} d^4x_1 d^4x_2 \ldots d^4x_n \qquad (6.18)$$

Thus,

$$S = I + i \int_{-\infty}^{\infty} H_{\zeta}^{\zeta}(x_1) d^4x_1 - \frac{1}{2!} \int_{-\infty}^{\infty} \int_{-\infty}^{\infty} T\left\{H_{\zeta}^{\zeta}(x_1) H_{\zeta}^{\zeta}(x_2)\right\} d^4x_1 d^4x_2 \ldots . \sum_{n=0}^{\infty} S^{(n)} \qquad (6.19)$$

We present Wick's theorem which is given such that (Klauber, 2013):

$$(6.20)$$

$$T\left\{(AB \ldots)_{x_1} \ldots (AB \ldots)_{x_n}\right\}$$
$$= N\left\{(AB \ldots)_{x_1} \ldots (AB \ldots)_{x_n}\right\} + N\left\{(AB \ldots)_{x_1} (AB)_{x_2} \ldots .\right\}$$
$$+ N\left\{\ldots (A \ldots . Z)_{x_{n-1}} (A \ldots . Z)_{x_n}\right\}$$

Where it will be conveyed that notations $AB \ldots Z$ and $H_{\zeta}^{\zeta}(x_1)$ pertain to particle field functions $\overline{\psi}$ and ψ at positions x_μ such that:

$$(AB \ldots)_{x_1} = H_{\zeta}^{\zeta}(x_1) = \left(\overline{\psi} \gamma^{\mu} A_\mu \psi\right)_{x_1} \qquad (6.21)$$

$$(AB \ldots)_{x_2} = H_{\zeta}^{\zeta}(x_2) = \left(\overline{\psi} \gamma^{\mu} A_\mu \psi\right)_{x_2} \qquad (6.22)$$

where the gamma matrices γ^{μ} have values such that (Klauber, 2013):

$$\gamma^0 = \begin{bmatrix} 1 & 0 & 0 & 0 \\ 0 & 1 & 0 & 0 \\ 0 & 0 & -1 & 0 \\ 0 & 0 & 0 & -1 \end{bmatrix} \qquad \gamma^1 = \begin{bmatrix} 0 & 0 & 0 & 1 \\ 0 & 0 & 1 & 0 \\ 0 & -1 & 0 & 0 \\ -1 & 0 & 0 & 0 \end{bmatrix} \qquad (6.23)$$

$$\gamma^2 = \begin{bmatrix} 0 & 0 & 0 & -i \\ 0 & 0 & i & 0 \\ 0 & i & 0 & 0 \\ -i & 0 & 0 & 0 \end{bmatrix} \qquad \gamma^3 = \begin{bmatrix} 0 & 0 & 1 & 0 \\ 0 & 0 & 0 & -1 \\ -1 & 0 & 0 & 0 \\ 0 & -1 & 0 & 0 \end{bmatrix}$$

Where $\bar{\psi}$ and ψ are particle/antiparticle fields and A_{μ} is an arbitrary vector quantity, the Dyson expansion is essentially fused with Wick's theorem. Hence, the S operator is expressed in terms of fields $\bar{\psi}$ and ψ and the gamma matrices γ^{μ} such that (Klauber, 2013):

$$(6.24)$$

$$S = I + ie \int_{-\infty}^{\infty} \left(\bar{\psi} \, \gamma^{\mu} A_{\mu} \psi \right)_{x_1} d^4 x_1$$

$$- \frac{e^2}{2!} \int_{-\infty}^{\infty} \int_{-\infty}^{\infty} T \left\{ \left(\bar{\psi} \, \gamma^{\mu} A_{\mu} \psi \right)_{x_1} \left(\bar{\psi} \, \gamma^{\mu} A_{\mu} \psi \right)_{x_2} \right\} d^4 x_1 d^4 x_2 + \cdots + \sum_{n=0}^{\infty} S^{(n)}$$

The values of $S^{(0)}$, $S^{(1)}$, and $S^{(2)}$ of the Dyson-Wick expansion are given such that (Klauber, 2013):

$$S^{(0)} = I \qquad (6.25)$$

$$S^{(1)} = ie \int_{-\infty}^{\infty} \left(\bar{\psi} \, \gamma^{\mu} A_{\mu} \psi \right)_{x_1} d^4 x_1 \qquad (6.26)$$

$$S^{(2)} = -\frac{e^2}{2!} \int_{-\infty}^{\infty} \int_{-\infty}^{\infty} T \left\{ \left(\bar{\psi} \, \gamma^{\mu} A_{\mu} \psi \right)_{x_1} \left(\bar{\psi} \, \gamma^{\mu} A_{\mu} \psi \right)_{x_2} \right\} d^4 x_1 d^4 x_2 \qquad (6.27)$$

Where I (or simply 1) is the multiplicative identity, the weak interaction between particle/antiparticle pairs will correlate to S operator $S^{(2)}$ for our purposes. The bra vector $\langle i|$ represents the initial state of the particle interaction and the ket vector $|f\rangle$ is the final state of the interaction. Hence, bra-ket notation expresses the transition amplitude such that (Klauber, 2013):

$$\langle i|S^{(2)}|f\rangle = S_{if} \tag{6.28}$$

Where S_{if} denotes the initial and final transition amplitude, the transition amplitude can be expressed such that (Klauber, 2013):

$$\langle i|S^{(2)}|f\rangle = S_{if} = -\frac{e^2}{2!}\int_{-\infty}^{\infty}\int_{-\infty}^{\infty} T\left\{\left(\bar{\psi}\gamma^{\mu}A_{\mu}\psi\right)_{x_1}\left(\bar{\psi}\gamma^{\mu}A_{\mu}\psi\right)_{x_2}\right\}d^4x_1 d^4x_2 \tag{6.29}$$

Where the initial state transpires at position x_1 and the final state transpires at position x_2, the probability of the interaction is given such that (Klauber, 2013):

$$S_{if}S_{if}^{\dagger} = |S_{if}|^2 \tag{6.30}$$

Consider the Feynman propagator in the wave number space (or $k_{\mu} \in R^n$) of (Klauber, 2013):

$$\Delta_F(x-y) = \frac{1}{(2\pi)^4}\int \frac{e^{-k_{\mu}(x_{\mu}-x'_{\mu})}}{(k_{\mu})^2-\mu^2+i\varepsilon}d^4k_{\mu} \tag{6.31}$$

Energy E_{med} is the energy of the particle mediating the weak interaction between the particle/antiparticle pair. Hence, we set energy E_{med} equal to relativistic energy such that (Klauber, 2013):

$$E_{med} = pc \tag{6.32}$$

Where p denotes momentum and c the velocity of light, momentum p takes on a value in terms of the inverse gravity concept of (Klauber, 2013):

$$p = \frac{E_{med}}{c} = \left[\frac{\propto_c}{cN_c r_0^2}\right]\left[\frac{(l_{med})^3}{3GM_A'M_B'}\right] \tag{6.33}$$

As expressed in chapter 5, a_μ is a directional unit vector, the vector valued momentum p_μ can be expressed as (Klauber, 2013):

$$p_\mu = pa_\mu \tag{6.34}$$

Therefore, vector valued wave number k_μ is expressed such that:

$$k_\mu = \frac{p_\mu}{\hbar} \tag{6.35}$$

Where $\hbar$ denotes reduced Planck's constant, wave number $k_{g\mu}$ denotes the wave number in terms of the inverse gravity concept. Inverse gravity wave number $k_{g\mu}$ takes a value such that:

$$k_{g\mu} = \frac{a_\mu}{\hbar}\left[\frac{\propto_c}{N_c r_0^2}\right]\left[\frac{(l_{med})^3}{3GM_A'M_B'}\right] \tag{6.36}$$

Thus, the Feynman propagator $\Delta_{IG}(x_\mu - x_\mu')$ in terms of inverse gravity wave number $k_{g\mu}$ is expressed such that (Klauber, 2013):

$$\Delta_{IG}(x_\mu - x_\mu') = \frac{1}{(2\pi)^4}\int\frac{e^{-k_{g\mu}(x_\mu - x_\mu')}}{(k_{g\mu})^2 - \mu^2 + i\varepsilon}d^4 k g_\mu \tag{6.37}$$

Where

$$0.001 \text{ fm} \geq l_{med} = |x_u' - x_u| \tag{6.38}$$

The propagator $\Delta_{IG}(x_\mu - x_\mu')$ in terms of the inverse gravity concept constitutes the mediating or virtual particle conducting, in this case, the weak interaction between particle/antiparticle pairs described by field functions $\psi^\dagger(x_\mu)$ and $\psi(x_\mu)$ (where $\bar\psi$ and $\psi \rightarrow \psi^\dagger(x_\mu)$ and $\psi(x_\mu)$). Thus, the exchange of a mediating virtual particle between particle/antiparticle pairs between positions x_1 and x_2 is given such that (Klauber, 2013):

$$S^{(2)} = -\frac{e^2}{2!} \int_{-\infty}^{\infty} \int_{-\infty}^{\infty} d^4x_1 d^4x_2 \psi^\dagger(x_1)\, \gamma^\text{u}\psi(x_1)\Delta_{IG}\left(x_\mu - x_\mu'\right)\psi^\dagger(x_2)\, \gamma^\text{u}\, \psi(x_2) \tag{6.39}$$

So alternatively,

$$S^{(2)} = -\frac{e^2}{2!} \int_{-\infty}^{\infty} \int_{-\infty}^{\infty} d^4x_1 d^4x_2 \psi^\dagger(x_1)\, \gamma^\text{u}\psi(x_1)\left[\frac{1}{(2\pi)^4}\int \frac{e^{-kg\mu(x_\mu - x_\mu')}}{(kg\mu)^2 - \mu^2 + i\varepsilon} d^4 kg_\mu\right]\psi^\dagger(x_2)\, \gamma^\text{u}\, \psi(x_2)$$
$$\tag{6.40}$$

Thus, the transition amplitude can be expressed in terms of bra-ket notation such that (Klauber, 2013):

$$\langle i|S^{(2)}|f\rangle =$$
$$-\frac{e^2}{2!} \int_{-\infty}^{\infty} \int_{-\infty}^{\infty} d^4x_1 d^4x_2 \psi^\dagger(x_1)\, \gamma^\text{u}\psi(x_1)\left[\frac{1}{(2\pi)^4}\int \frac{e^{-kg\mu(x_\mu - x_\mu')}}{(kg\mu)^2 - \mu^2 + i\varepsilon} d^4 kg_\mu\right]\psi^\dagger(x_2)\, \gamma^\text{u}\, \psi(x_2)$$
$$\tag{6.41}$$

Or in terms of positions x_1 and x_2, this can be expressed as:

$$\langle x_1|S^{(2)}|x_2\rangle =$$
$$-\frac{e^2}{2!} \int_{-\infty}^{\infty} \int_{-\infty}^{\infty} d^4x_1 d^4x_2 \psi^\dagger(x_1)\, \gamma^\text{u}\psi(x_1)\left[\frac{1}{(2\pi)^4}\int \frac{e^{-kg\mu(x_\mu - x_\mu')}}{(kg\mu)^2 - \mu^2 + i\varepsilon} d^4 kg_\mu\right]\psi^\dagger(x_2)\, \gamma^\text{u}\, \psi(x_2)$$
$$\tag{6.42}$$

Where $\psi^\dagger(x_\mu)$ and $\psi(x_\mu)$ corresponds to position x_1 and $\psi^\dagger(x_\mu')$ and $\psi(x_\mu')$ and corresponds to position x_2, the transition amplitude adheres to the Feynman diagram of figure 4 shown below.

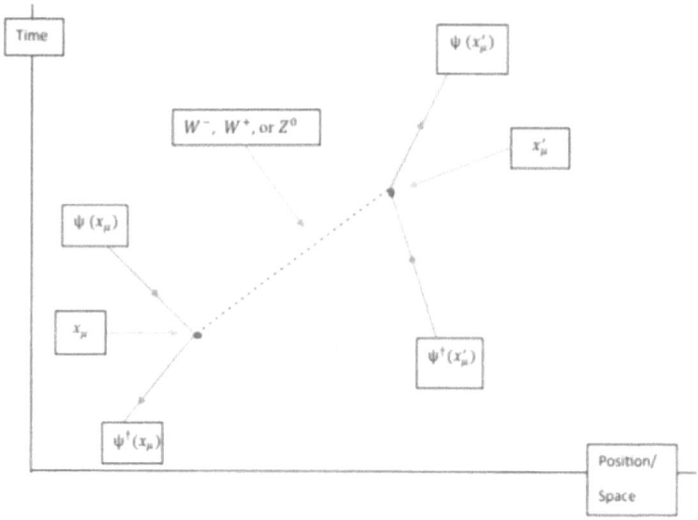

Figure 4

Hence, the particle field $\psi(x_\mu)$ and its antiparticle field $\psi^\dagger\left(x_\mu\right)$ are expressed in terms of ½ spin particles such that (Klauber, 2013):

$$\psi(x_\mu) = \int \frac{d^3 p}{\sqrt{2(2\pi)^3}} \bar{u}_f(p)\hat{a}^\dagger e^{i\frac{p_\mu x_\mu}{\hbar}} + \int \frac{d^3 p}{\sqrt{2(2\pi)^3}} \bar{v}_f(p)\hat{b} e^{-i\frac{p_\mu x_\mu}{\hbar}} \tag{6.43}$$

Or alternatively,

$$\psi = \psi^+ + \psi^- \tag{6.44}$$

And (Klauber, 2013),

$$\psi^\dagger\left(x_\mu\right) = \int \frac{d^3 p}{\sqrt{2(2\pi)^3}} \bar{u}_f(p)\hat{b}^\dagger e^{i\frac{p_\mu x_\mu}{\hbar}} + \int \frac{d^3 p}{\sqrt{2(2\pi)^3}} \bar{v}_f(p)\hat{a} e^{-i\frac{p_\mu x_\mu}{\hbar}} \tag{6.45}$$

Or alternatively,

$$\psi^\dagger = \psi^{\dagger -} + \psi^{\dagger +} \tag{6.46}$$

These fields have the form of the scalar field $\psi(x_\mu)$. The continuous scalar particle field $\psi(x_\mu)$ describing each ½ spin particle within the field is of the general form (Klauber, 2013):

$$\psi(x_\mu) = \int \frac{d^3p}{\sqrt{2(2\pi)^3}} \bar{u}_f(p)\hat{a}^\dagger e^{i\frac{p_\mu x_\mu}{\hbar}} + \int \frac{d^3p}{\sqrt{2(2\pi)^3}} \bar{v}_f(p)\hat{b}e^{-i\frac{p_\mu x_\mu}{\hbar}} \tag{6.47}$$

where the plane wave function ($e^{i\frac{p_\mu x_\mu}{\hbar}}$) for free particles are at momentum and positions values p_μ and x_μ ($p_\mu x_\mu$) in R^3 or three-dimensional space respectively. The integration element d^3p corresponding to the momentum space has the form such that:

$$d^3p = \prod_1^3 dp_\mu \tag{6.48}$$

The spinors of $\bar{u}_f(p)$ and $\bar{v}_f(p)$ correspond to the ½ spin fermion and the ½ spin antiparticle in the form of scalar field $(\psi(x_\mu))$. Thus, the values of spinors $\bar{u}_f(p)$ and $\bar{v}_f(p)$ correspond to the solution set of (Klauber, 2013):

$$\bar{u}_1(p) = \sqrt{\frac{E+m_p}{2m_p}} \begin{bmatrix} 1 \\ 0 \\ \frac{p_3}{E+m_p} \\ \frac{p_1+ip_2}{E+m_p} \end{bmatrix} \qquad \bar{u}_2(p) = \sqrt{\frac{E+m_p}{2m_p}} \begin{bmatrix} 0 \\ 1 \\ \frac{p_1-ip_2}{E+m_p} \\ -\frac{p_3}{E+m_p} \end{bmatrix} \tag{6.49}$$

$$\bar{v}_1(p) = \sqrt{\frac{E+m_p}{2m_p}} \begin{bmatrix} \frac{p_3}{E+m_p} \\ \frac{p_1+ip_2}{E+m_p} \\ 1 \\ 0 \end{bmatrix} \qquad \bar{v}_2(p) = \sqrt{\frac{E+m_p}{2m_p}} \begin{bmatrix} \frac{p_1-ip_2}{E+m_p} \\ -\frac{p_3}{E+m_p} \\ 1 \\ 0 \end{bmatrix} \tag{6.50}$$

As expressed in the seminal paper, p_1, p_2, and p_3 are the momentum values in 3-space (e.g., $p_u = (p_1, p_2, p_3)$) corresponding to the momentum operator of $p_u = i\hbar\partial/\partial x_u$; m_p is the particle rest mass and E is the corresponding energy value. Momentum p_u and energy E obey the relativistic energy equation of (Klauber, 2013):

$$E^2 = p_1{}^2 c^2 + p_2{}^2 c^2 + p_3{}^2 c^2 + m_p{}^2 c^4 \tag{6.51}$$

Thus, the products of spinors $\bar{u}_f(p)$ and $\bar{v}_f(p)$ (for indices $f = f$) obey the orthogonal relations of (Klauber, 2013):

$$\bar{u}_1(p)\bar{u}_1(p) = \frac{E}{m_p} \qquad \bar{u}_2(p)\bar{u}_2(p) = \frac{E}{m_p} \tag{6.52}$$

$$\bar{v}_1(p)\bar{v}_1(p) = \frac{E}{m_p} \qquad \bar{v}_2(p)\bar{v}_2(p) = \frac{E}{m_p} \tag{6.53}$$

And thus, for the condition of indices such that $f \neq g$, we have (Klauber, 2013):

$$\bar{u}_1(p)\bar{u}_2(p) = 0 \qquad \bar{v}_1(p)\bar{v}_2(p) = 0 \tag{6.54}$$

which are of the form:

$$\bar{v}_f(p)\bar{u}_g(p) = 0 \qquad \bar{u}_f(p)\bar{v}_g(p) = 0 \tag{6.55}$$

Since the particles are created or destroyed (or annihilated) across interaction distance l_{med}, each particle (protons, neutrinos, and so on) exist with part of the total energy E_{med} of the mediating particle. Energy E presented within the solution set of spinors $\bar{u}_f(p)$ and $\bar{v}_f(p)$ is equal to the energy of the inverse gravity mediating particle of energy E_{med} as shown below.

$$E = E_{med} = \left[\frac{\alpha_c}{N_c r_0^2}\right]\left[\frac{(l_{med})^3}{3GM_A' M_B'}\right] \tag{6.56}$$

And thus,

78

$$\left[\left[\frac{\propto_c}{N_c r_0^2} \right] \left[\frac{(l_{med})^3}{3GM'_A M'_B} \right] \right]^2 = p_1^2 c^2 + p_2^2 c^2 + p_3^2 c^2 + m_p^2 c^4 \qquad (6.57)$$

Therefore, the products of spinors $\bar{u}_f(p)$ and $\bar{v}_f(p)$ (for $= f$) have values in terms of energy E_{med} such that (Klauber, 2013):

$$\bar{u}_f(p)\bar{u}_f(p) = \frac{1}{m_p} \left[\frac{\propto_c}{N_c r_0^2} \right] \left[\frac{(l_{med})^3}{3GM'_A M'_B} \right] \qquad (6.58)$$

$$\bar{v}_f(p)\bar{v}_f(p) = \frac{1}{m_p} \left[\frac{\propto_c}{N_c r_0^2} \right] \left[\frac{(l_{med})^3}{3GM'_A M'_B} \right] \qquad (6.59)$$

Therefore, in defining the components of the field functions $\psi^\dagger(x_\mu)$ and $\psi(x_\mu)$, the creation operators $(\hat{a}^\dagger, \hat{b}^\dagger)$ and the annihilation operators $(\hat{a}, \hat{b})$ of field $\psi(x_u)$ obey the antisymmetric property required for the description of fermions in accordance to the Pauli exclusion principle. The creation and annihilation operators have the properties such that (Klauber, 2013):

$$\{\hat{a}, \hat{a}^\dagger\} = \hat{a}\hat{a}^\dagger + \hat{a}^\dagger\hat{a} = \delta \qquad\qquad \{\hat{b}, \hat{b}^\dagger\} = \hat{b}\hat{b}^\dagger + \hat{b}^\dagger\hat{b} = \delta \qquad (6.60)$$

$$\{\hat{a}^\dagger, \hat{a}^\dagger\} = \{\hat{a}, \hat{a}\} = 0 \qquad\qquad \{\hat{b}^\dagger, \hat{b}^\dagger\} = \{\hat{b}, \hat{b}\} = 0 \qquad (6.61)$$

And

$$\hat{a}^\dagger\hat{a} = \hat{b}^\dagger\hat{b} = n \qquad (6.62)$$

Where n is the number of particles in the system and δ is the Kronecker delta which has the orthonormal property of (Klauber, 2013):

$$\delta_j^i = \begin{cases} i \neq j, \delta = 0 \\ i = j, \delta = 1 \end{cases} \qquad (6.63)$$

This completes Description 1.

Description 2: The Hamiltonian of the beta decay process incorporating the inverse gravity concept

As expressed in my paper titled "Cosmology: The Theoretical Possibility of Inverse Gravity as a Cause of Cosmological Inflation in an Isotropic and Homogenous Universe and Its Relationship to Weakly Interacting Massive Particles," we give an example of a weak interaction being conducted by an inverse gravity mediating particle. Thus, as expressed in the seminal paper, we derive the Feynman, Gell-Mann, Marshak, and Sudarshan constructions for describing the beta decay process. The continuous scalar fields of each elementary particle are denoted as the neutrino field $n(x_u)$, the positron field $p(x_u)$, the electron or beta minus particle field $\beta^-(x_u)$, and the antineutrino field $\bar{v}_e(x_u)$ corresponding to the beta decay process $(n \rightarrow p + \beta^- + \bar{v}_e)$ within Fermi's theory which are of the previously introduced form of (Walker, 2017):

$$\psi(x_u) = \int \frac{d^3p}{\sqrt{2(2\pi)^3}} \bar{u}_f(p)\hat{a}^\dagger e^{i\frac{p_u x_u}{\hbar}} + \int \frac{d^3p}{\sqrt{2(2\pi)^3}} \bar{v}_f(p)\hat{b} e^{-i\frac{p_u x_u}{\hbar}} \tag{6.64}$$

And

$$\psi^\dagger(x_u) = \int \frac{d^3p}{\sqrt{2(2\pi)^3}} \bar{u}_f(p)\hat{b}^\dagger e^{i\frac{p_u x_u}{\hbar}} + \int \frac{d^3p}{\sqrt{2(2\pi)^3}} \bar{v}_f(p)\hat{a} e^{-i\frac{p_u x_u}{\hbar}} \tag{6.65}$$

We present the current-by-current Hamiltonian $H^{Int}(x'_u, x_u)$ of Fermi's theorem of the beta decay process for the interaction $(n \rightarrow p + \beta^- + \bar{v}_e)$. Therefore, this constitutes the current-by-current Hamiltonian $H^{Int}(x'_u, x_u)$ of the beta decay process at positions x'_u and x_u (over interaction distance l_{med}). I must specify that the Hamiltonian $H^{Int}(x'_u, x_u)$ only pertains to the energy of the weak interaction within the nucleus of an atom. So I posit that inverse gravity can occur within the nucleus of an atom, which was not sufficiently expounded upon in the seminal paper; however, the inverse gravity effect is negligible at this level. Although the effects of inverse gravity are negligible at the subatomic level within a nucleus, I still proceed to conduct the analysis of an inverse gravity weak interaction of Description 2. Moreover, this conveys that inverse gravity energy can occur anywhere in space. Hence, the Hamiltonian $H^{Int}(x'_\mu, x_\mu)$ of the weak interaction is expressed such that (Casalbuoni, 1997):

$$H^{Int}(x'_\mu, x_\mu) = \frac{G_F}{\sqrt{2}} J^\mu(x_\mu) J_\mu(x'_\mu) \tag{6.66}$$

Where $J^\mu(x_\mu)$ denotes the current density corresponding to particle fields $\psi^\dagger_{AL,R}(x_\mu)$ and $\psi_{BL,R}(x_\mu)$ at position x_μ and $J_\mu(x'_\mu)$ denotes the current density corresponding to fields $\psi^\dagger_{CL,R}(x'_\mu)$ and $\psi_{DL,R}(x'_\mu)$ at position x'_μ. Current densities $J^\mu(x_\mu)$ and $J_\mu(x'_\mu)$ at positions x_μ and x'_μ have values such that (Casalbuoni, 1997):

$$J^\mu(x_\mu) = \left(\psi^\dagger_{AL,R}(x_\mu)\gamma^\mu\psi_{BL,R}(x_\mu)\right) \qquad J_\mu(x'_\mu) = \left(\psi^\dagger_{CL,R}(x'_\mu)\gamma_\mu\psi_{DL,R}(x'_\mu)\right) \tag{6.67}$$

The subscripts L, R denote the left- and right-handed projections of the particle fields which will be defined with more detail shortly; the subscripts $A, B, C,$ and D denote the type of particle field (e.g., protons, electrons, neutrons, etc.). The particle fields $\psi^\dagger_{AL,R}(x_\mu)$ and $\psi_{BL,R}(x_\mu)$ at position x_μ of current density $J^\mu(x_\mu)$ relate to the particle fields $\psi^\dagger_{CL,R}(x'_\mu)$ and $\psi_{DL,R}(x'_\mu)$ at position x'_μ of current density $J_\mu(x'_\mu)$ via the mediating particle of energy E_{med} over distance l_{med} of the weak interaction. The general expression of the corresponding Hamiltonian $H^{Int}(x'_\mu, x_\mu)$ is given such that (Casalbuoni, 1997):

$$H^{Int}(x'_\mu, x_\mu) = \frac{G_F}{\sqrt{2}} \left(\psi^\dagger_{AL,R}(x_\mu)\gamma^\mu\psi_{BL,R}(x_\mu)\right)\left(\psi^\dagger_{CL,R}(x'_\mu)\gamma_\mu\psi_{DL,R}(x'_\mu)\right) + h.c. \tag{6.68}$$

The notation $h.c.$ denotes "Hermitian conjugation", G_F denotes Fermi's constant which has a value of 1.166×10^{-5} GeV^{-2}. Where $\gamma^\mu = \eta\gamma_\mu$ and γ_μ is the gamma matrices, η is the Minkowski metric, the field functions $\psi^\dagger_{AL,R}(x_\mu)$, $\psi_{BL,R}(x_\mu)$, $\psi^\dagger_{CL,R}(x'_\mu)$, and $\psi_{DL,R}(x'_\mu)$ are the right- and left-handed projections previously mentioned. Just a side note: in the original papers, I used the notation of Φ to denote ½ spin particle fields, which was a mistake being that the physics community commonly denotes ½ spin fields with ψ. Hence, in continuing the derivation, it is well known that weak interactions do not obey the parity law of symmetry. Thus, the characteristic of chirality within the beta decay process states that the right-handed field signifies that the momentum and spin transpire in the same direction while the left-handed field signifies that the momentum and spin have opposite directions; therefore, the right- and left-handed field functions are of the form (Casalbuoni, 1997):

$$\psi_R(x_u) = \psi(x_u)\frac{1+\gamma^5}{2} \qquad\qquad \psi_L(x_u) = \psi(x_u)\frac{1-\gamma^5}{2} \tag{6.69}$$

Where $\frac{1+\gamma^5}{2}$ and $\frac{1-\gamma^5}{2}$ are the right- and left-hand operators respectively. Either field function can be presented in the general form (Casalbuoni, 1997) :

$$\psi_{R,L}(x_u) = \psi(x_u)\frac{1\pm\gamma^5}{2} \tag{6.70}$$

The value γ^5 or gamma 5 matrix has a value such that (Casalbuoni, 1997):

$$\gamma^5 = -\frac{\vec{p}\cdot\vec{\Sigma}}{|\vec{p}|} \tag{6.71}$$

where $\vec{p}$ is a momentum vector expressed such that (Walker, 2017):

$$\vec{p} = (mc, p_1, p_2, p_3) \equiv p_u \tag{6.72}$$

where the modulus of momentum vector $\vec{p}$ is denoted $|\vec{p}|$ has a value such that (Walker, 2017):

$$|\vec{p}| = [(mc)^2 + (p_1)^2 + (p_2)^2 + (p_3)^2]^{1/2} \tag{6.73}$$

The vector quantity $\vec{\Sigma}$ containing matrix valued components has values such that (Walker, 2017):

$$\vec{\Sigma} = \begin{bmatrix} \bar{\sigma} & \mathbf{0} \\ \mathbf{0} & \bar{\sigma} \end{bmatrix} = \left\{ \begin{bmatrix} 0 & 0 \\ 0 & 0 \end{bmatrix}, \begin{bmatrix} \sigma_x & \mathbf{0} \\ \mathbf{0} & \sigma_x \end{bmatrix}, \begin{bmatrix} \sigma_y & \mathbf{0} \\ \mathbf{0} & \sigma_y \end{bmatrix}, \begin{bmatrix} \sigma_z & \mathbf{0} \\ \mathbf{0} & \sigma_z \end{bmatrix} \right\} \tag{6.74}$$

where

$$\vec{p} \bullet \vec{\Sigma} = p_1 \begin{bmatrix} \sigma_x & \mathbf{0} \\ \mathbf{0} & \sigma_x \end{bmatrix} + p_2 \begin{bmatrix} \sigma_y & \mathbf{0} \\ \mathbf{0} & \sigma_y \end{bmatrix} + p_3 \begin{bmatrix} \sigma_z & \mathbf{0} \\ \mathbf{0} & \sigma_z \end{bmatrix} \tag{6.75}$$

where bold zero expressed as $\mathbf{0}$ denotes the 4 by 4 zero matrices such that (Walker, 2017):

$$\mathbf{0} = \begin{bmatrix} 0 & 0 \\ 0 & 0 \end{bmatrix} \tag{6.76}$$

The symbol $\bar{\sigma}$ conventionally denotes the 2 by 2 Pauli spin matrices $\bar{\sigma}$ such that (Walker, 2017):

$$\bar{\sigma} \quad \rightarrow \quad \sigma_x = \begin{pmatrix} 0 & 1 \\ 1 & 0 \end{pmatrix}, \quad \sigma_y = \begin{pmatrix} 0 & -i \\ i & 0 \end{pmatrix}, \quad \sigma_z = \begin{pmatrix} 1 & 0 \\ 0 & -1 \end{pmatrix} \tag{6.77}$$

The matrices $\vec{\Sigma}$ obey the relation of (Walker, 2017):

$$\bar{\alpha} = \gamma^5 \vec{\Sigma} \tag{6.78}$$

And matrices $\bar{\alpha}$ take on values of (Walker, 2017):

$$\bar{\alpha} = \begin{bmatrix} \mathbf{0} & \bar{\sigma} \\ \bar{\sigma} & \mathbf{0} \end{bmatrix} = \left\{ \begin{bmatrix} \mathbf{0} & \sigma_x \\ \sigma_x & \mathbf{0} \end{bmatrix}, \begin{bmatrix} \mathbf{0} & \sigma_y \\ \sigma_y & \mathbf{0} \end{bmatrix}, \begin{bmatrix} \mathbf{0} & \sigma_z \\ \sigma_z & \mathbf{0} \end{bmatrix} \right\} \tag{6.79}$$

Now that we have defined the components of the system, the Hamiltonian $H^{Int}(x'_u, x_u)$ of the beta decay process ($n \rightarrow p + \beta^- + \bar{\nu}_e$) can be expressed in terms of the neutrino field $n_{L,R}(x_u)$, the positron field $p_{L,R}(x_u)$, the electron or beta minus particle field $\beta^-{}_{L,R}(x'_u)$, and the antineutrino field $\bar{\nu}_{e\,L,R}(x'_u)$ such that (Casalbuoni, 1997):

$$H^{Int}(x'_u, x_u) = \frac{G_F}{\sqrt{2}} \left(n_{L,R}(x_u) \gamma^u p_{L,R}(x_u) \right) \left(\beta^-{}_{L,R}(x'_u) \gamma_u \bar{\nu}_{e\,L,R}(x'_u) \right) + h.c. \tag{6.80}$$

Hence, the particle field functions $n_{L,R}(x_u)$, $p_{L,R}(x_u)$, $\beta^-{}_{L,R}(x'_u)$, and $\bar{v}_{e\,L,R}(x'_u)$ are of the form of $\psi_{R,L}(x_u)$ and replace the notation of the particle fields of $\psi^\dagger_{AL,R}(x_u)$, $\psi_{BL,R}(x_u)$, $\psi^\dagger_{CL,R}(x'_u)$, and $\psi_{DL,R}(x'_u)$. The weak interactions energy E_{med} will be denoted $E(x'_u, x_u)$. We revert to the notation of $|x'_u - x_u|$ for interaction length l_{med} in the expression of weak interaction energy E_{med} shown below (Walker, 2017).

$$E(x'_u, x_u) = E_{med} = \left[\frac{\alpha_c}{N_c r_0^2}\right]\left[\frac{|x'_u - x_u|^3}{3GM'_A M'_B}\right] \tag{6.81}$$

Thus, within the interactions $(n \rightarrow p + \beta^- + \bar{v}_e)$ of the particle field functions $n_{L,R}(x_u)$, $p_{L,R}(x_u)$, $\beta^-{}_{L,R}(x'_u)$, and $\bar{v}_{e\,L,R}(x'_u)$, the constituent particles annihilate one another producing a virtual particle (or mediating particle) of inverse gravity energy $E(x'_u, x_u)$. In turn, this creates another set of particles. The Hamiltonian $H^{Int}(x'_u, x_u)$ accounts for the energy of the entire system and thus interaction; this implies that the Hamiltonian $H^{Int}(x'_u, x_u)$ of the beta decay process equals energy $E(x'_u, x_u)$ as shown below (Walker, 2017).

$$\Delta H^{Int}(x'_u, x_u) = \Delta E(x'_u, x_u) \tag{6.82}$$

This completes Description 2.

Description 3: The Lagrange of an electroweak interaction incorporating the inverse gravity cosmological concept

The weak interactions of dark matter, which may in theory drive cosmological expansion, may have roots within electroweak interactions; a unified description of electromagnetic interactions and weak interactions is expressed in the introduction. Thus, I will express the inverse gravity concept in terms of electroweak interactions. I begin with ψ_L and ψ_R which are the left- and right-handed wave functions of the form previously introduced and shown below.

$$\psi_R(x_\mu) = \psi(x_\mu)\frac{1+\gamma^5}{2} \qquad\qquad \psi_L(x_\mu) = \psi(x_\mu)\frac{1-\gamma^5}{2} \qquad (6.83)$$

Thus, we will begin with gauge transformations of $SU(2) \times U(1)$ which are (Devanathan, 2019):

$$\psi_L \rightarrow \psi'_L = e^{i\alpha\cdot T + i\beta(Y/2)}\psi_L \qquad (6.84)$$

$$\psi_R \rightarrow \psi'_R = e^{i\beta(Y/2)}\psi_R \qquad (6.85)$$

where ψ_L and ψ_R are the left- and right-handed wave functions and ψ'_L and ψ'_R constitute the transformation under $SU(2) \times U(1)$. Variables α and β will be independent of space-time coordinates and for local phase invariance. Variable T denotes isospin (and thus the third or z component of spin), and Y denotes hypercharge ($Y = strangeness + baryon\ number$). The equation of charge Q is expressed such that (Devanathan, 2019):

$$Q = T + \frac{Y}{2} \qquad (6.86)$$

Consider the Dirac Lagrangian of the electron/neutrino pair $L(a)$ (where $a = \{\psi_L, \psi_R, \psi'_L, \psi'_R\}$), which is expressed as (Devanathan, 2019):

$$L(a) = \overline{\psi}_L \gamma^\mu (i\partial_\mu)\psi_L + \overline{\psi}_R \gamma^\mu (i\partial_\mu)\psi_R \qquad (6.87)$$

which is invariant under gauge transformation $SU(2) \times U(1)$. This means that:

$$L(\psi_L, \psi_R) = L(\psi'_L, \psi'_R) \qquad (6.88)$$

Showing symmetry between values $L(\psi_L, \psi_R)$ and $L(\psi'_L, \psi'_R)$, the Lagrange $L(\psi'_L, \psi'_R)$ can be expressed such that (Devanathan, 2019):

$$L(\psi'_L, \psi'_R) = \overline{\psi}'_L \gamma^\mu (i\partial_\mu)\psi'_L + \overline{\psi}'_R \gamma^\mu (i\partial_\mu)\psi'_R \qquad (6.89)$$

As conveyed by Devanathan, for four vector gauge bosons W_1, W_2, W_3, and B, we introduce a left- and right-handed covariant derivative D_μ which will be minimally substituted for the left- and right-hand partial derivative ∂_μ such that (Devanathan, 2019):

$$(i\partial_\mu)_R \rightarrow (iD_\mu)_R \tag{6.90}$$

$$(i\partial_\mu)_L \rightarrow (iD_\mu)_L \tag{6.91}$$

Therefore,

$$(iD_\mu)_L = (i\partial_\mu)_L - gTW - g'\frac{1}{2}YB_\mu \qquad Y = -1 \tag{6.92}$$

$$(iD_\mu)_R = (i\partial_\mu)_R - g'\frac{1}{2}YB_\mu \qquad Y = -2 \tag{6.93}$$

where the strength of the SU(2) coupling to the Gauge fields is denoted by g. Substituting the covariant derivatives $(iD_\mu)_L$ and $(iD_\mu)_R$ into the Lagrange $L(\psi_L, \psi_R)$ gives (Devanathan, 2019):

$$L(\psi_L, \psi_R) = \overline{\psi}_L \gamma^\mu \left((i\partial_\mu)_L - gTW - g'\frac{1}{2}YB_\mu \right) \psi_L + \overline{\psi}_R \gamma^\mu \left((i\partial_\mu)_R - g'\frac{1}{2}YB_\mu \right) \psi_R \tag{6.94}$$

To represent kinetic energies and self-coupling of the W_μ fields and the kinetic energy field B_μ, we transform Lagrange $L(\psi_L, \psi_R)$ such that (Devanathan, 2019):

$$L(\psi_L, \psi_R) \rightarrow L' = L(\psi_L, \psi_R) - \frac{1}{4}W_{\mu\nu}W^{\mu\nu} - \frac{1}{4}B_{\mu\nu}B^{\mu\nu} \tag{6.95}$$

Where $W^{\mu\nu}$ and $B^{\mu\nu}$ are the contravariant matrices to $W_{\mu\nu}$ and $B_{\mu\nu}$, Lagrange L' takes on a value such that:

$$L' = L(\psi_L, \psi_R) - \frac{1}{4}W_{\mu\nu}W^{\mu\nu} - \frac{1}{4}B_{\mu\nu}B^{\mu\nu} \tag{6.96}$$

where the values of $W_{\mu\nu}$ and $B_{\mu\nu}$—corresponding to kinetic energies and self-coupling of the W_μ fields and the kinetic energy field B_μ—are expressed such that (Devanathan, 2019):

$$W_{\mu\nu} = \partial_\mu W_\nu - \partial_\nu W_\mu - gW_\mu \times W_\nu \tag{6.97}$$

$$B_{\mu\nu} = \partial_\mu B_\nu - \partial_\nu B_\mu \tag{6.98}$$

which are of the form of the field strength tensor $F_{\mu\nu}$ of:

$$F_{\mu\nu} = \partial_\mu A_\nu - \partial_\nu A_\mu \tag{6.99}$$

which in turn are equivalent to the 4 by 4 matrix such that:

$$F_{\mu\nu} = \begin{bmatrix} 0 & -E_x/c & -E_y/c & -E_z/c \\ E_x/c & 0 & B_z & -B_y \\ E_y/c & -B_z & 0 & B_x \\ E_z/c & B_y & -B_x & 0 \end{bmatrix} \tag{6.100}$$

where components E_a and B_a are components of the electric field (E) and magnetic field (B) in three-dimensional space (or x, y, and z) respectively, and c is the velocity of light. Thus, the Eigen values corresponding to covariant derivatives $(iD_\mu)_R$ and $(iD_\mu)_L$ correlating to particle fields ψ_L and ψ_R are expressed such that (Devanathan, 2019):

$$(iD_\mu)_L \psi_L = \left[(i\partial_\mu)_L - gT \cdot W - g'\frac{1}{2}YB_\mu \right] \psi_L \tag{6.101}$$

$$(iD_\mu)_R \psi_R = \left[(i\partial_\mu)_R - g'\frac{1}{2}YB_\mu \right] \psi_R \tag{6.102}$$

At this juncture, we can make the correlation to the inverse gravity concept through the equivalence of (Devanathan, 2019):

$$(iD_\mu)_L \psi_L = (i\partial_\mu)_L \psi_L = k_{g\mu,L} \psi_L \tag{6.103}$$

87

$$\left(iD_\mu\right)_R \Psi_R = \left(i\partial_\mu\right)_R \Psi_R = k_{g\mu,R} \Psi_R \qquad (6.104)$$

Where the left- and right-handed inverse gravity wave numbers $k_{g\mu L}$ and $k_{g\mu R}$ each have a value such that:

$$k_{g\mu L} = \frac{a_\mu}{\hbar}\left[\frac{\propto_{cL}}{N_c r_0^2}\right]\left[\frac{(l_{med})^3}{3GM_A' M_B'}\right] \qquad (6.105)$$

$$k_{g\mu R} = \frac{a_\mu}{\hbar}\left[\frac{\propto_{cR}}{N_c r_0^2}\right]\left[\frac{(l_{med})^3}{3GM_A' M_B'}\right] \qquad (6.106)$$

The proportionality constants pertaining to the left- and right-handed wave numbers $k_{g\mu L}$ and $k_{g\mu R}$ are $\propto_{cL}$ and $\propto_{cR}$ respectively. Hence, the Eigen vector and Eigen values for particle fields Ψ_L and Ψ_R in terms of the inverse gravity concept are given by (Devanathan, 2019):

$$\left[\left(i\partial_\mu\right)_L - gT \cdot W - g'\frac{1}{2}YB_\mu\right]\Psi_L = \left[\frac{a_\mu}{\hbar}\left[\frac{\propto_{cL}}{N_c r_0^2}\right]\left[\frac{(l_{med})^3}{3GM_A' M_B'}\right]\right]\Psi_L \qquad (6.107)$$

$$\left[\left(i\partial_\mu\right)_R - g'\frac{1}{2}YB_\mu\right]\Psi_R = \left[\frac{a_\mu}{\hbar}\left[\frac{\propto_{cR}}{N_c r_0^2}\right]\left[\frac{(l_{med})^3}{3GM_A' M_B'}\right]\right]\Psi_R \qquad (6.108)$$

which implies the equivalence of (Devanathan, 2019):

$$\left(iD_\mu\right)_L \Psi_L = \left[\frac{a_\mu}{\hbar}\left[\frac{\propto_{cL}}{N_c r_0^2}\right]\left[\frac{(l_{med})^3}{3GM_A' M_B'}\right]\right]\Psi_L \qquad (6.109)$$

$$\left(iD_\mu\right)_R \Psi_R = \left[\frac{a_\mu}{\hbar}\left[\frac{\propto_{cR}}{N_c r_0^2}\right]\left[\frac{(l_{med})^3}{3GM_A' M_B'}\right]\right]\Psi_R \qquad (6.110)$$

Therefore, the value of $L(\Psi_L', \Psi_R')$ of the Lagrange within total Lagrange L' can be expressed in terms of the inverse gravity concept such that:

$$L(\psi'_L, \psi'_R) = \overline{\psi}'_L \gamma^{\mu} \left[\frac{a_{\mu}}{\hbar} \left[\frac{\propto_{cR}}{N_c r_0^2} \right] \left[\frac{(l_{med})^3}{3GM'_A M'_B} \right] \right] \psi'_L + \overline{\psi}'_R \gamma^{\mu} \left[\frac{a_{\mu}}{\hbar} \left[\frac{\propto_{cL}}{N_c r_0^2} \right] \left[\frac{(l_{med})^3}{3GM'_A M'_B} \right] \right] \psi'_R \qquad (6.111)$$

Recall that (Devanathan, 2019):

$$L' = L(\psi_L, \psi_R) - \frac{1}{4} W_{\mu\nu} W^{\mu\nu} - \frac{1}{4} B_{\mu\nu} B^{\mu\nu} \qquad (6.112)$$

Thus, the total Lagrange L' can be expressed in terms of the inverse gravity concept such that:

$$(6.113)$$

$$L' = \overline{\psi}'_L \gamma^{\mu} \left[\frac{a_{\mu}}{\hbar} \left[\frac{\propto_{cL}}{N_c r_0^2} \right] \left[\frac{(l_{med})^3}{3GM'_A M'_B} \right] \right] \psi'_L + \overline{\psi}'_R \gamma^{\mu} \left[\frac{a_{\mu}}{\hbar} \left[\frac{\propto_{cR}}{N_c r_0^2} \right] \left[\frac{(l_{med})^3}{3GM'_A M'_B} \right] \right] \psi'_R -$$
$$\frac{1}{4} W_{\mu\nu} W^{\mu\nu} - \frac{1}{4} B_{\mu\nu} B^{\mu\nu}$$

This completes Description 3 and thus a holistic description of the inverse gravity cosmological concept as applied to weak interactions that constitute weakly interacting massive particles. I hope that you enjoyed reading my ideas and research of the inverse gravity cosmological concept.

Acknowledgements

I would like to thank and dedicate this book to my Lord and Savior Jesus Christ. Additionally, I would like to acknowledge Juanita Walker, my mother; Lola Mae Roberts, my grandmother; my father, Edward Walker III; my grandfather, Edward Walker II; my sister, Brittani Walker-Shingles; my brother, Kenneth Jonathan Oliver; and my nephews, Robert Edward Shingles and Logan R. Thompson. Lastly, I would like to give a special thanks to Mr. and Mrs. Eugene Thompson for their support and encouragement.

Bibliography

Casalbuoni, R. (1997). The Standard Model of Electroweak Interactions. Retrieved from
http://theory.fi.infn.it/casalbuoni/corso.pdf

Devanathan, V. (2019) How do the Particles Acquire Mass? The Gauge Theories, Higgs Field
and Higgs Bosons. E-Journal of Chennai Academy of Sciences 1, 1–20 .

Dobrijevic, D., and E. Howell (2023). *James Webb Space Telescope (JWST) — A complete
guide.* Space.com. https://www.space.com/21925-james-webb-space-telescope-jwst.html

Gregersen, Erik (March 20, 2023). Weakly interacting massive particles. Britannica.
https://www.britannica.com/science/weakly-interacting-massive-particle

Hirata, Christopher M. (November 5, 2012). Lecture VIII: Linearized gravity, Caltech M/C 350-17, Pasadena CA 91125, USA*. Retrieved from http://www.tapir.caltech.edu/~chirata/ph236/lec08.pdf

Hirvonenin, V. (n.d.). The Friedmann Equations Explained: A Complete Guide. Profound physics. https://profoundphysics.com/the-friedmann-equations-explained-a-complete-guide/

Klauber, R. D. (2013). *Student Friendly Quantum Field Theory*, Chapter 4, Sandtrove Press.

Liboff, R. L. (2003). *Introductory Quantum mechanics* (4th ed.). Addison Wesley.

Ojeda, P., and H. Rosu (2006). Supersymmetry of FRW barotropic cosmologies, *International Journal of Theoretical Physics*, **45**, no. 6, 1152–1157. https://doi.org/10.1007/s10773-006-9123-2

Wald, R. M. (1984). *General Relativity*, Chicago Press, Ltd., Chicago, Illinois. https://doi.org/10.7208/chicago/9780226870373.001.0001

Walker, E. A. (2017). Cosmology: The Theoretical Possibility of Inverse Gravity as a Cause of Cosmological Inflation in an Isotropic and Homogeneous Universe and its Relationship to Weakly

Interacting Massive Particles, *Advanced Studies in Theoretical Physics*, **11**, no. 11, 487-518. Retrieved from: https://doi.org/10.12988/astp.2017.7730

Walker, E. A. (2018). The Relationship between the Cosmological Inverse Gravity Assertion and the Cosmological Constant Including an Alternative Possibility to the Graviton. *Advanced Studies in Theoretical Physics*, *12*(1), 37-55.

_Winther, H. A. (Retrieved 2023). A course on the formation of the cosmic microwave background and structures in the Universe :Learn the theory, get the physical understanding and make your own CMB code. Institute for Theoretical Astrophysics at the University of Oslo. Cosmology II: Observables (wintherscoming.no)

Young, H. D., and R.A. Freedman (2004). *Sears and Zemansky's University Physics*, 11th Edition, Addison Wesley, San Francisco, Cal., Pearson, 2004.

Edward A. Walker is a minister, teacher, adjunct professor, and physics researcher. He currently resides in South Florida.

Index

www.ingramcontent.com/pod-product-compliance
Lightning Source LLC
Chambersburg PA
CBHW021449210526

45463CB00002B/692